The Southern African *Plectranthus*

Plate 1

Plectranthus **'Mona Lavender'**

(*P. hilliardiae* 'Magwa' × *saccatus*)

Half-title page:

Plectranthus ciliatus

THE SOUTHERN AFRICAN

Plectranthus

and the art of turning shade to glade

Ernst van Jaarsveld

Illustrations by Vicki Thomas

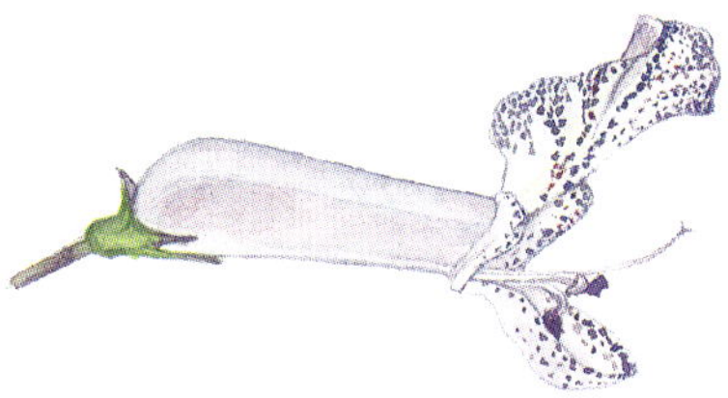

FERNWOOD
PRESS

THE SPONSOR

The publisher gratefully acknowledges the financial contribution made by the Cape Tercentenary Foundation to the publication of this book.

FERNWOOD PRESS

P O Box 481, Simon's Town 7995, South Africa
www.fernwoodpress.co.za

Registration no. 2005/166094/23
First published 2006

Historical photographs were acquired from the following sources:
The Linnean Society of London: Francis Masson, p.13;
The Royal Botanic Gardens, Kew, United Kingdom: P.J. Redouté, *Plectranthus fruticosus,* p.12;
George Bentham, p.14; Johann Drège, p.15.
The publisher acknowledges permission to reproduce these illustrations, with thanks.

Edited by Barbara Knox-Shaw
Production management by Antonia Venturini
Designed by Bruce Henderson
Cartography and typesetting by Daniele Michelini
Dustjacket design by Bruce Henderson

Reproduction by Hirt & Carter, Cape Town
Printed and bound by Tien Wah Press (Pte) Ltd, Singapore

ISBN 978-1-874950-80-6

Contents

Plate 2

Plectranthus hereroensis

Origin: Witpoortjie, Witwatersrand, Gauteng (486/77)

Foreword

Dr L. E. Codd

Although it has been realized for a number of years that many of our *Plectranthus* species would make attractive garden subjects, it is only recently, since Ernst van Jaarsveld took an interest in them, that gardeners in general are becoming aware of them. Displays held at Kirstenbosch during the past few years, organized initially by Ernst and later by other members of the Kirstenbosch staff, have demonstrated many species growing to perfection, either in massed shrub plantings or as groundcovers in shady places. In addition, several species were shown to be well suited for growing as ornamental pot plants in hanging baskets.

Since his appointment to the staff of the National Botanical Gardens of South Africa, stationed first at Nelspruit from 1974 to 1976 and then at Kirstenbosch, Ernst has made a significant contribution to our knowledge of the genus by collecting specimens throughout South Africa. It is obvious that he has an eagle eye for spotting unusual plants and in the process has discovered three new species, one of which I had the pleasure of naming in his honour. His collections have also added greatly to our awareness of the extent to which species can vary, whilst several appear to be interspecific hybrids. The fact that he has regularly collected living material as well has greatly added to the value of his contributions.

My own interest in the group goes back over thirty years when I found that it was almost impossible to make a reliable identification of a *Plectranthus* specimen in the herbarium. This was partly because the plants are difficult to turn into good herbarium specimens, with the result that the early attempts at classification often increased the confusion. Fortunately the original collecting sites of most species are on record, and the ease with which plants can be propagated by cuttings greatly facilitated their study in cultivation.

In addition to its horticultural interest, *Plectranthus* shows a number of fascinating features which are worthy of closer study. The vegetative parts are often characteristically scented, from the rich, pleasant fragrance of *P. amboinicus* and the citronella-like scent of *P. laxiflorus* to the strong and rather repellent odour of *P. neochilus*.

As constituted at present, the genus forms a somewhat heterogeneous group of about 300 species whose distribution extends from Africa and Madagascar, where the main concentrations are, to the southern Arabian peninsula and India, with a few outliers in Malesia and Australia. Preliminary cytological studies have revealed that several series exist within the genus, each differing in basic chromosome numbers, and it is possible that the whole group will eventually be subdivided into several genera. However, such a step would be unwise until the numerous tropical African species have been thoroughly sorted out and their affinities with the southern African species established.

L. E. Codd
Botanical Research Institute
Pretoria
1986

This Foreword by the late Dr Codd, a former Director of the Botanical Research Institute and author of the last revision of the South African Plectranthus, *was written in 1986 when the first version of this book was completed.*

Plate 3
Plectranthus verticillatus
'Pink Surprise'

For Erma

Acknowledgements

Many people have contributed to the making of this book, and I am grateful to all the following: to Vicki Thomas for the painting of all the various species and for her patience; Trevor Edwards who initially translated and edited parts of the text; Roger Jacques who also edited some of the text and prepared the glossary; Barbara Knox-Shaw who edited the final version; and Piet Vorster who edited the original Afrikaans text. During the early years I received help and inspiration from Leslie Codd, Elize Buitendach and Jo Onderstall, while help in the field came from various farmers and colleagues including Tony Abbott, Elaine Campher, David Culverwell, Paul Ems, Anthony Hitchcock, Johan Kluge, Hugh Nicholson, Geoff Nicholls, Andrew Hankey, Dave Hardy, David Styles, Jan Burring, Wessel Swanepoel, Pieter Winter, Liesl van der Walt, Fanie Venter, and Werner Voigt. I am also grateful to John Winter for allowing so many of the *Plectranthus* field trips, and to Tommy Sardien, Andrew Jacobs, Eric Marthinus and Trevor Solomons for propagating and cultivating the large number of species at Kirstenbosch. Gert Brits is especially thanked for freely sharing information on the plectranthus he bred, as well as for letting me use his photographs, and allowing me to take some of my own as well. My greatest thanks go to Erma, my wife and life partner, for her never-failing encouragement and understanding, her help on many expeditions, and her patience with my many absences in the field, and to whom this book is dedicated.

By the same author

The Plectranthus Handbook (1987)
A Revision of the Genus Gasteria (1994)
Wonderful Waterwise Gardening (2000)
Vygies, Gems of the Veld (with U. de Villiers Pienaar) (2000)
Succulents of South Africa (with Ben-Eric van Wyk
and Gideon Smith) (2000)
Cotyledon and Tylecodon (with Daryl Koutnik) (2004)
Contributor (with U. de Villiers Pienaar) to *Les Mésembs d'Afrique du Sud* (2004)

Introduction

South Africa is well known for its floral diversity (about 23 000 species), and its historic contribution to world ornamental horticulture has been great. For years the two top-selling houseplants gracing the balconies of Europe, the United States and Japan have been pelargonium cultivars and the hen-and-chickens, *Chlorophytum comosum*. Others that have achieved international fame are the proteas, aloes, mesembs, clivias, gladioli, freesias, Barberton daisies and the arum or calla lilies. There remains a large reservoir of wild unexploited species, and some of these are beginning to attract the attention of gardeners and plant breeders here and abroad. Among them is the previously neglected group of *Plectranthus*, also known as the spurflowers, and the recent release of two stunning new cultivars, the mauve 'Mona Lavender' and the pink 'Cape Angel', bred from *Plectranthus hilliardiae* and *P. saccatus*, marks the dawning of a new era in indigenous plant breeding.

Plectranthus is an Old World genus belonging to the Mint family (Lamiaceae), and there are about 350 known species which are widely distributed in Africa, occurring also in India and Australia (even Japan). Of these, 53 occur in South Africa and Namibia, but it is in the Eastern Cape and KwaZulu-Natal regions where the species with the greatest horticultural potential lie.

The Mint family, which is best represented in the Mediterranean region, is full of gardenworthy plants. In South Africa it includes 36 genera, with about 274 species. *Plectranthus* is the largest genus (53), with *Stachys* (41) second. Other genera that show horticultural potential include *Salvia* (23), *Hemizygia* (28), *Leonotis* (3), *Syncolostemon* (10), *Pycnostachys* (3), *Tetradenia* (5), *Thorncroftia* (5) and *Aeollanthus* (8).

The name *Plectranthus* means spurflower (*plectron* = spur and *anthos* = flower) and refers to the characteristic spur at the base of the corolla tube of *Plectranthus fruticosus*, the first plectranthus to be placed in the genus. The name is confusing because very few plectranthus actually have this spur, but the French botanist L'Héritier, who described the genus in 1788, did not know this.

The South African plectranthus have long been known in Europe, and especially in Scandinavia, where they are known as 'Swedish Ivy', and are commonly planted in windowboxes and hanging baskets. Their popularity there and in the United States has increased greatly in the last few years, following the appearance of a number of new cultivars and hybrids, some of which are even sold in the supermarkets.

What make plectranthus so appealing? They have ornamental leaves, make free-flowering, easy and fast-growing plants, and resist most pests and diseases. They make colourful displays in the autumn when other flowers are often scarce. They tolerate fairly heavy shade, and as groundcovers they suppress weed growth, prevent erosion and allow leaf litter to accumulate, which helps the soil retain moisture. In the wild, they are the succulent plants of the forests. Like *Pelargonium*, their succulent nature – which makes them such good house plants – is an adaptation to disturbance from animals and to drought. Their ability to root means that they can recover from damage so that a dropped branch, for example, can survive. They tolerate mishandling, and can build confidence in gardeners who think they do not have green fingers. Even when they are desiccated by drought they recover rapidly after a light rain shower. Their shallow root system and tolerance of root competition also counts in their favour as garden plants. Almost all species are rapid growers, most layer abundantly and form luxuriant mats or dense stands. Most are suitable as container subjects for hanging baskets or as bushy patio plants.

The value of plectranthus as garden plants has long been recognized. The botanist Edwin Phillips, writing in 1928, acknowledged that 'the South African species certainly deserve more attention from horticulturists than has been given', adding that they were well worth cultivating and made 'very fine ornamental shade plants'. More recently the garden writer David Hessayon, writing about plectranthus in Sweden in 1980, pointed out that three different species were called 'Swedish Ivy', and commented, 'It is a pity that Swedish Ivy is not more popular in other countries as it has many good points'. Despite the popularity of these few species and the 'many good points' of the group, relatively few plectranthus are widely cultivated and much horticultural potential still lies untapped.

The Southern African plectranthus have been under close horticultural scrutiny at Kirstenbosch Botanical Garden since 1974 when I became interested in the genus. Since then I have collected, propagated, cultivated, evaluated and described many different plectranthus, and tried to make them better known to botanists and gardeners. Many are now grown in the National Botanical Institute's eight National Botanic Gardens, and they are increasingly being used in private and public gardens, and sold as container plants. Some selection and breeding have given rise to several new cultivars and hybrids. As these gain popularity, I hope that there will be greater recognition of the group's potential, and that some of the lesser-known species will also become more widely cultivated.

I first completed this book in Afrikaans in the late seventies but could not find a publisher at the time. Trevor Edwards from the University of KwaZulu-Natal initially translated and edited parts of the text, and I later finished the translation and revised it, bringing it up to date. The taxonomic account follows the revision by Dr. L. E. Codd for the *Flora of Southern Africa* (1986), and all subsequent findings – whether they are new species or reclassifications – have been incorporated.

All the indigenous and naturalized *Plectranthus* of South Africa and Nambia are described here, and are illustrated by Vicki Thomas. I have also included a few other species from Mt Mulanje in Malawi that grow well in South Africa.

All accession numbers refer to Kirstenbosch unless otherwise stated.

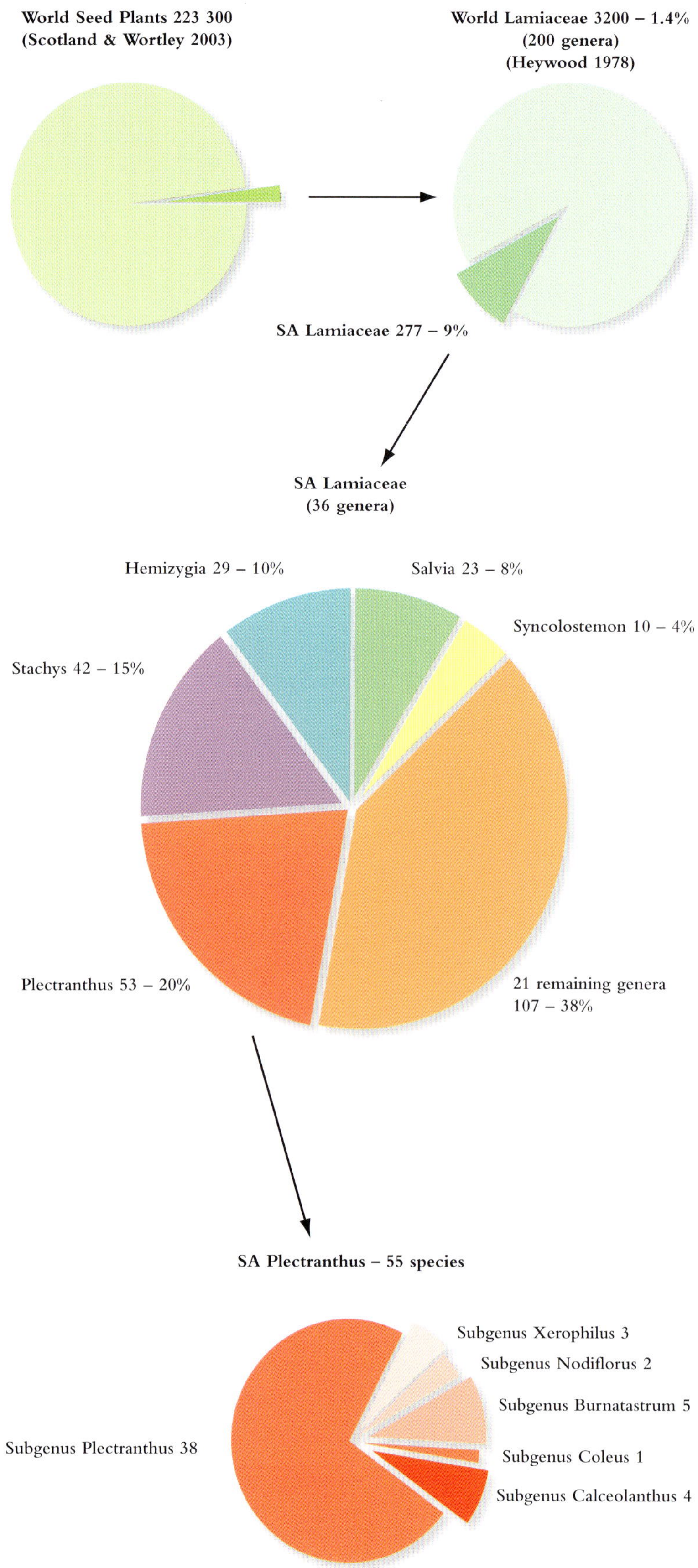

***Plectranthus* in relation to seed plants and other Lamiaceae**

Plate 4 ***Plectranthus ecklonii***
The common blue form, widespread in forests from the Eastern Cape to Barberton in Mpumalanga

XLI.

P. J. Redouté del. Maleuvre sculp.

PLECTRANTHUS fruticosus.

1 The history of *Plectranthus*

The first introductions of Cape plants to Europe caused tremendous excitement, particularly the bold proteas, aloes and cycads which seized the early travellers' attention. *Plectranthus*, however, remained unnoticed until well into the eighteenth century. These more humble shrubs and herbs are from the summer-rainfall forests and savannahs, and in the prime collecting area of the winter-rainfall Cape fynbos they are either absent or rare. This probably explains why they do not appear in herbaria before the seventeenth century, nor are they mentioned in Carl Linnaeus' famous *Species Plantarum* of 1753.

Although they were not known to botanists, one *Plectranthus* in particular, *P. esculentus* (umbondwe or African potato) would have been widely grown in the warm sandy African savannahs. This starch-rich potato-like crop is still planted today, and there has been renewed interest in its use as a drought-resistant food source. It was documented by Dr N. E. Brown from a Medley Wood specimen in the *Kew Bulletin* of 1894.

Plectranthus hadiensis

The first *Plectranthus* to be described was *P. hadiensis*, which was discovered by Pehr Forsskål in Hadiyah in the Yemen, though it also occurs in South Africa. In his 1775 flora of Egypt and Arabia, Forsskål named it *Ocimum hadiense*, placing it in the closely related sweet basil genus that had been established by Carl Linnaeus in 1753. There it remained, unnoticed by botanists, until 1894 when Schweinfurth and Sprenger re-examined the type specimen and transferred it to *Plectranthus*.

Kew, Carl Thunberg and Francis Masson

Towards the end of the eighteenth century several prominent botanists visited the Cape, rekindling interest in South African plants including *Plectranthus*.

Carl Thunberg (1743–1828), a pupil of Linnaeus, is known as the 'Father of South African Botany'. He was a Swedish botanist who arrived at the Cape in April 1772. He made three major journeys inland, on the second of which he was accompanied by Francis Masson (1741–1805), the young Scottish horticulturalist who was collecting for the recently established Royal Botanic Gardens at Kew. The director of Kew was the great naturalist Sir Joseph Banks, who had visited the Cape briefly in 1771 while on Captain Cook's *Endeavour* expedition, and what he saw made such an impression on him that he determined to send a collector to gather seeds for Kew.

During his first visit to the Cape (1772–1775) Masson undertook two major collecting trips, and it was on the second, which lasted from September 1773 to January 1774 that he and Thunberg travelled to the southern Cape as far east as the Sundays River. On this journey they collected three new *Plectranthus* species: *P. verticillatus*, *P. fruticosus* and *P. madagascariensis*. *P. verticillatus*, collected by Thunberg in the Outeniqua mountains, was described by the younger Linnaeus as *Ocimum verticillatum* in his *Species Plantarum* 3 of 1781, the second *Plectranthus* to be described. Although Linnaeus believed that the type specimen had been collected by Montin in India, we now know that Montin received his plants from Thunberg (Codd 1975, 408). Thunberg

Francis Masson, the plant collector sent by Kew to the Cape in 1772 and again in 1786. He probably introduced more than 1500 Cape plants to Europe

Opposite: The earliest illustration of *Plectranthus fruticosus* by the famous flower painter, P. J. Redouté, from L'Héritier's *Stirpes Novae* of 1784. The plant illustrated was raised from seed collected by Masson on his expedition to the Eastern Cape with Carl Thunberg in 1774

must thus be credited with the discovery of *P. verticillatus*. To add to the complexity, when the botanist George Bentham (1800–1884) revised the now established *Plectranthus* genus in 1832, he (wrongly) named it *P. thunbergii*, unaware that the younger Linnaeus had already established the name *Ocimum verticillatum* from Thunberg's specimen. The plant was finally named *Plectranthus verticillatus* by George Druce in 1917.

In the meantime William Aiton, chief horticulturist at Kew, succeeded in raising *P. fruticosus* from Masson's seed. In the *Hortus Kewensis* of 1780 (II, 3) he listed it along with two other non-South Africans (*P. punctatus*, and *P. barbatus* which was naturalized in but not indigenous to South Africa), so *P. fruticosus* was also the first of the South African plectranthus to be taken up as an indoor plant in Europe.

Founding of the genus *Plectranthus*

The genus *Plectranthus* was founded in 1788 by a French magistrate and keen botanist, Charles Louis L'Héritier de Brutelle (1746–1800), the son of a nobleman. In 1786, while working on a South American collection made by the French botanist Joseph Dombey (1742–1793), L'Héritier came to hear that the Spanish Government wanted to claim Dombey's collection for Spain. L'Héritier immediately left for England taking the collection with him, and there he was befriended by Sir Joseph Banks, who helped him continue his work of describing new species without hindrance. When he saw the as yet unnamed *P. fruticosus* which William Aiton had raised from Masson's seed at Kew, he realized that it belonged to a new unnamed genus. Basing his description both on Masson's new plant, which has a spur at the base of its corolla, as well as on *P. punctatus*, he named the genus *Plectranthus* (*plectron* = spur and *anthos* = flower). What L'Héritier did not realize was that this spur is absent from most other *Plectranthus* species, but once a name is validly published it has to remain. To accompany his description of the new genus he commissioned an illustration from his protégé, Pierre-Joseph Redouté, whom he had persuaded to join him in London. Redouté's plate together with the new description were published in L'Héritier's *Stirpes Novae* of March 1788 (see p.12). This, the third formally described South African species, was therefore the first to be described as a *Plectranthus*.

Competition among botanists is ever-present. Jean Baptiste Lamarck (1744–1829), another French botanist, also recognized *P. fruticosus* as a new genus, and in April 1788, one month after L'Héritier, he described the new genus *Germanea* with two species, *G. urticifolia* (which corresponded to L'Héritier's *P. fruticosus*) and *G. maculosa*. According to the rules of international botanical nomenclature, however, the earliest validly published name must stand, so *Germanea* became the first synonym of *Plectranthus*.

George Bentham trained as a lawyer, but an inheritance enabled him to devote his time to botany. He was responsible for the first major revision of *Plectranthus* in 1876

The genus *Coleus* and *Plectranthus amboinicus*

Another early genus, *Coleus* (now incorporated within *Plectranthus*) was described by the Portuguese missionary and naturalist João de Loureiro (1717–1791). He established his new genus on the basis of the stamens which are fused at the base and not free as in *Plectranthus*. He based his type on *Coleus* (now *Plectranthus*) *amboinicus*, another species of southern African origin, and the fourth *Plectranthus* species to be named. It was transferred to *Plectranthus* in 1825, though Dr Leslie Codd, in his 1975 revision, did recognize *Coleus* as a subgenus within *Plectranthus*.

Loureiro's type specimen, which is in a poor state, has since been discarded in favour of one collected by Arthur F. G. Kerr (1877–1944) at Siam in the Far East. The story of how this African plant came to be common in the East was told by Dr Codd (1975):

> *Plectranthus amboinicus* has become widely distributed by cultivation, first in the Far East and, later, in the West Indies and tropical America. Its strong, not unpleasant aroma has led to its use medicinally and as a popular spice for flavoring food, being known as Soup Mint French Thyme, Spanish Thyme, Country Borage and Indian Mint. According to Trimen it is employed in Ceylon as a medicine especially for cattle and a plant is always to be found growing in a little box suspended on the side of the native carts. There has been speculation as to its country of origin but it undoubtedly occurs naturally in Africa from Kenya southwards to Angola and Natal. It is common near Lourenco Marques (e.g. Inhaca Island where it would have been accessible to the early voyagers to the east).

This easily grown succulent trailer is therefore one of the first to have been cultivated, and its medicinal value helped it to spread rapidly.

George Bentham's revisions of *Plectranthus*

George Bentham (1800–1884) was a leading British taxonomist, and the author of many important publications and floras. He was also secretary of the Horticultural Society from 1829–40 and president of the Linnaean Society from 1861–74. His first account of *Plectranthus* appeared in 1832 in his *Labiatarum Genera et Species* (1832–6), where he recognized seven sections: *Germanea*, *Coleoides*, *Heterocylix*, *Melissoides*,

Isodon, *Pyramidium* and *Amethystoides*. He continued to work on *Plectranthus*, writing the section 'Labiatae' for Alphonse De Candolle's *Prodromus* in 1848, and in 1876 he revised the genus further in the major work he wrote with Sir Joseph Hooker, *Genera Plantarum* (1862–83). Here he recognized only two main sections, *Germanea* (with two subsections, *Germanea* and *Coleoides*) and *Isodon* (with four subsections, *Isodon*, *Pyramidium*, *Amethystoides* and *Melissoides*). *Germanea* here contains most of the mainstream *Plectranthus* species, the chief characteristics of which are the sessile cymes and the 2-lipped calyces each with a solitary broad-toothed upper lip and four narrow acuminate teeth. His section *Isodon* is characterized by species with evenly toothed calyces carried on pedunculate cymes.

Nineteenth-century discoveries and their collectors

With the expansion of the British Empire and the founding of the Royal Botanic Gardens at Kew, a major wave of travelling botanists reached the Cape, and 29 names covering most of the widespread species were added in the nineteenth century.

Although *Plectranthus madagascariensis* was collected by Thunberg and Masson, it was only described in 1806 under the name *Ocimum madagascariensis* by Christiaan Persoon (1762–1836) from plants collected by the French botanist Philibert Commerson (1727–1775) in Madagascar. This was transferred to *Plectranthus* in 1832. *P. barbatus* followed shortly, described from seed collected by Lord Valentia (1769–1844) in Abyssinia in 1809. The third South African *Plectranthus* named in the nineteenth century was the widespread annual, *P. caninus*, though it was actually collected in India (where it also occurs) by the Moravian missionary Benjamin Heyne (1770–1819), and it was named by the botanist John Rotheram (1750–1804), a pupil of Linnaeus, in 1821.

William John Burchell (1781–1863), a naturalist and traveller from London, collected during his time at the Cape more than 8000 plant specimens, among which is the first record of *P. ecklonii* in the Eastern Cape. Burchell was also responsible for collecting *P. lucidus* (or *P. strigosus* var. *lucidus* as Bentham named it in 1848; it was raised to specific status in 1997).

Johann Drège (1794–1881), a German horticulturist and botanical collector, came to the Cape in 1826. Drège was one of the greatest botanical explorers of his time. During his eight-year visit to South Africa he travelled extensively in the Eastern and Western Cape and collected more than 200 000 specimens. Among these were seven new *Plectranthus* species named by the botanist Bentham. *P. saccatus* and *P. ciliatus* were described in 1837 from plants Drège collected at the Mzimvubu River near Port St Johns; *P. ambiguus* (then named *P. coloratus*) was collected between the Mzimvubu and Mgazana Rivers; *P. laxiflorus* Benth. between the Mzimkulu and Umkomaas; and *P. petiolaris* between the Umtata and Mzimvubu. The last of his discoveries, *P. spicatus*, came from Glenfilling in the Eastern Cape. Drège was also the first to collect *P. hadiensis* var. *tomentosus* in KwaZulu-Natal, which George Bentham named in 1837. Although Dr Codd commemorated him in *P. dregei* in 1957, this was later found to have an earlier name, *P. ambiguus*. He is remembered, however, in *Euphorbia dregeana* and *Oxalis dregei*, amongst others.

Christian Ecklon (1795–1868) and Carl Zeyher (1799–1858) were two German botanists who travelled and worked together in South Africa. Ecklon collected *P. strigosus* (named by Bentham in 1837) at the Olifantshoek Forest near Alexandria in the Eastern Cape. Although Burchell was the first to collect *P. ecklonii*, Bentham named this species after Ecklon in 1848, basing the description on his Katberg collection.

Wilhelm Schimper (1804–1878) was a German collector who travelled widely in Arabia and north Africa. He collected the widespread succulent *P. cylindraceus* at Samen, near Gapdia in Ethiopia. This was named by the German botanist Christian Hochstetter (1787–1860) in 1848.

Joseph Burke (1812–1873), an Englishman from Bristol, was a collector of natural history specimens for Lord Derby. He arrived at the Cape in 1840 and left in 1842 after extensive travels in the north-eastern part of the country. He collected *P. ramosior* in the Magaliesberg, and this was described by George Bentham in 1848.

William Harvey (1811–1866) was Colonial Treasurer at the Cape from 1836 to 1842. He named *P. purpuratus* in 1859 from seed sent to Kew from Port Natal by Mr Vause.

Hermann Wilhelm Rudolf Marloth (1855–1931) was a trained pharmacist and botanist from Lubben in Germany who settled at the Cape in December 1883. He travelled widely in South Africa and Namibia and collected *P. hereroensis* in 1886. It was named by Heinrich Engler (1844–1930) in 1888 from plants collected by Marloth on the Wilhelmsberg near Okahandja.

Alexander Mechow collected a shrubby *Plectranthus* during his visit to Angola (1879–1881) somewhere between the Malanga and Cuango Rivers. This was named by John Briquet (1870–1931) in 1894 as *Coleus mirabilis*, and transferred to *Plectranthus* by Edmund Launert (1926–) in 1968.

John Medley Wood (1827–1915) was a retired British naval officer who settled in KwaZulu-Natal and later became curator of the Botanical Gardens in

Johann Drège came to the Cape in 1826 from Germany, and stayed until 1834. He travelled widely, collecting hundreds of thousands of specimens. He discovered seven new *Plectranthus* species

Durban where he grew a number of plectranthus. The type specimen of *P. esculentus* described by Dr N.E. Brown in 1894 was based on a plant specimen collected and grown by Medley Wood. He also grew *P. hadiensis* var. *tomentosus*, *P. ambiguus*, *P. saccatus* subsp. *saccatus* and *P. ecklonii*. In his *Natal Plants*, Medley Wood reported that *P. saccatus* was already popular as a garden plant by the end of the nineteenth century in Durban. He sent seed of this species to the Cambridge University Botanical Gardens in 1888, and one of the plants raised was illustrated in *Curtis's Botanical Magazine* in 1902. The tuberous-rooted *P. woodii* was named by the botanist Robert Gürke (1854–1911) in his honour in 1898.

Carl Hugo Holst (1865–1892) collected the first specimens of the yellow-flowering annual *P. tetragonus* at Usambara, Mashena in Tanzania. It was named by Gürke in the *Botanischer Jahrbuch* (1894).

P. ornatus, a popular garden plant in South Africa, was wrongly named *P. comosus* by Christian Hochstetter in 1894. It is a Kenyan species, later renamed by Dr Codd in 1975.

Peter MacOwan (1830–1909), a teacher and botanist, collected a *Plectranthus* species near Grahamstown which Harry Bolus (1834–1911) named *Orthosiphon ambiguus* in 1881. Dr Leslie Codd realized what it was and correctly transferred it to *Plectranthus* in 1964. In fact the plant had already been named *Plectranthus coloratus* by E. Meyer from a specimen collected by Drège, but this could not be upheld because the name had already been used for a different species of *Plectranthus*.

Ernest Edwin Galpin (1858–1941) was a widely travelled amateur botanist from Grahamstown who made a considerable contribution to the knowledge of the South African flora. He was the first to collect *P. neochilus* at Rimers Creek in Mpumalanga. It was named by Rudolph Schlechter (1872–1925) in 1896.

William Tyson (1851–1920) was the son of missionary parents living in Jamaica. He settled in South Africa and was a teacher by profession. A keen naturalist, he collected *P. grandidentatus* on the Enyembe Mountain in east Griqualand in the Eastern Cape. It was named by Gürke in 1898.

Anton Rehmann (1840–1917) was an Austrian botanist and explorer. He visited South Africa on two occasions, in 1875–7 and again in 1879–80. He collected about 9000 plant specimens of which several were new. These included four new *Plectranthus* species: *P. woodii* from Beggarsberg, described by Gürke in 1898; *P. rehmannii* which Gürke named in his honour; and *P. dolichopodus* and *P. elegantulus*, both named by John Briquet in 1903. The last three species were collected in Karkloof near Pietermaritzburg. Rehmann is also commemorated in the related *Aeollanthus rehmannii*.

John William Carnegie Kirk (1878–1962) was a British soldier and natural scientist who served widely in South and East Africa where he collected a number of plant specimens. *Coleus tetensis*, which he collected near Tete in Mozambique, was named by E. G. Baker in 1900. It was transferred to *Plectranthus* by the botanist A. D. Q. Agnew (1929–) in 1974. It is thus the last plectranthus found in the nineteenth century.

The mountains of the Blyde River Canyon, habitat of *Plectranthus rubropunctatus*, named by Dr Leslie Codd in 1974

Briquet's and Baker's revisions of 1897 and 1900

John Isaac Briquet (1870–1931) was a well-known Swiss botanist, and the 'Rapporteur-Général' for botanical nomenclature from 1900 to 1931 at Geneva. He was the author of many publications and scientific articles. He included all the *Plectranthus* known up to 1897 in his revision of the Labiatae in *Pflanzenfamilie*, No. 4. His classification is similar to that of Bentham, but he raised Bentham's sections *Germanea* and *Isodon* to subgenera. He also established Bentham's section *Isodon* as a new genus, *Burnatastrum*, mainly on the grounds of its circinate calyx and the flowers which are carried in a dichasium (*P. spicatus* is a South African member). Dr Codd recognized *Burnatastrum*, but reduced it to a section of *Plectranthus*.

Edmund G. Baker (1864–1949) revised *Plectranthus* for the *Flora of Tropical Africa* (vol. 5) in 1900, treating 74 species. His work is in line with Briquet's.

The twentieth century and the establishment of the Botanical Research Institute and the Kirstenbosch National Botanical Gardens

In the twentieth century botanical exploration was organized by the newly established universities, research institutes and botanical gardens, most notably Kirstenbosch which was founded in 1913 on the eastern slopes of Table Mountain with Professor Harold

Pearson as its first director and J. W. Matthews as the first curator. The cool moist air and high precipitation of its afrotemperate climate favoured the introduction of *Plectranthus*. In its first two years four of the most attractive and useful of the *Plectranthus* were introduced: *P. ecklonii* (1790/14), *P. fruticosus* (1841/14), *P. ciliatus* (3789/14) and *P. madagascariensis* (4869/14). By 1920 there were ten species some of which had been donated by Mr Carnegie, a chemist from Pietermaritzburg. These included *P. hadiensis* var. *tomentosus* (1091/17), *P. laxiflorus* (849/20), *P. dolichopodus* (1360/34) and *P. rehmannii*.

In 1903 the Botanical Research Institute was founded in Pretoria with Dr Burtt Davy as first director, and the Botanical Gardens attached to the Institute were established in 1946. Dr Leslie Codd (1908–1999) became director in 1963 and remained so until his retirement in 1973. The Institute was amalgamated with Kirstenbosch and all other National Botanical Gardens in 1989 to form the National Botanical Institute, which became known as the South African National Biodiversity Institute in 2005.

Dr Leslie Codd's revisions of 1975 and 1985

Although a treatment of all the South African *Plectranthus* by T. Cooke appeared in *Flora Capensis* of 1912, it was Dr Codd's major revision of the South African *Plectranthus* in *Bothalia* 11 of 1975 that laid a solid foundation. He took a broad approach to this highly variable genus by lumping together a number of smaller, closely related genera (*Burnatastrum* and *Coleus*) under the umbrella of *Plectranthus*. These together with the subgenera *Nodiflorus*, *Calceolanthus* and *Xerophilus* (named by Codd himself) may in the future be raised to independent genera, though most botanists agree that the present arrangement should be retained until more evidence such as DNA analysis will provide has been collected and studied.

Dr Codd was also responsible for naming ten new *Plectranthus* species and a new variety. These include *P. unguentarius*, *P. hilliardiae*, *P. mutabilis*, *P. rubropunctatus*, *P. psammophilus*, *P. saccatus* var. *longitubus*, *P. xerophilus*, *P. oribiensis*, *P. ernstii*, *P. praetermissus* and *P. dolomiticus*. His monograph was updated again for the *Flora of Southern Africa* in 1985.

Twentieth-century discoveries and collectors

In the twentieth century another 23 *Plectranthus* were discovered, including a number of interesting new species with limited distributions towards the close of the century.

Kurt Dinter (1868–1945), a German botanist and botanical explorer, was a great lover of succulents. He became curator of the famous garden La Mortola on the Italian Riviera, but left in 1897 to settle in Namibia. He organized many expeditions there and collected *P. dinteri* on the Waterberg, which Briquet named in his honour in 1903.

Rudolph Schlechter (1872–1925) was a German botanical collector who moved to South Africa in 1891. He worked briefly for the Department of Agriculture and also for Ralph Arderne who had a beautiful garden in Claremont, now known as the Arderne Gardens. Harry Bolus took him on as an assistant in the herbarium and library. It was during a collecting trip to KwaZulu-Natal and the north in 1893–4 that he collected *P. grallatus* at Mount Frere, which was named by John Briquet in 1903.

Arnold Penther (1855–1931) was born in Rome, and worked as a zoologist at the Natural History Museum in Vienna. During a collecting trip to South Africa and Zimbabwe in 1894 he travelled eastwards with Schlechter to Port Elizabeth, where he met up with P. Krook, a local plant collector, and together they gathered many plants for Penther's herbarium. One of their findings was *P. pentheri* (named *Coleus pentheri* by Gürke in 1905, transferred to *Plectranthus* in 1997) which Krook collected at Breakfast Vlei in the Eastern Cape. This species was sunk under *P. neochilus* by Codd in his revision of 1975, but its distinctly succulent roots and white flowers separate it immediately from that species so it has again been raised to species level. (Dr Codd was probably unaware of its distinctly tuberous roots as most specimens are prepared from aerial parts of the plant.)

William Gerrard (d.1866) came from Liverpool, and was a keen naturalist. He collected widely in KwaZulu-Natal from 1860 onwards, discovering many new species. He suffered much from malaria and moved to Madagascar in 1865. He collected the first specimens of *P. zuluensis* which was named by Thomas Cook (1869–1947) in the *Kew Bulletin* of 1909.

Charles Swynnerton (1877–1938) was a naturalist born in India. He came to KwaZulu-Natal in 1897, and moved to Melsetter in Zimbabwe in 1898 where he became manager of Gungunyana Farm in the Chipinga District. He collected *P. swynnertonii* in the Chirinda Forest in 1898. It was named by S. Moore (1851–1931) in the *Journal of the Linnaean Society* in 1911.

P. oertendahlii was named by the botanist Thore Fries (1886–1930) in 1924 from a cultivated plant in Sweden. This plectranthus had been grown for a considerable time in Sweden and other parts of Europe, but its wild origin remained a mystery until its location was discovered in southern KwaZulu-Natal (see pp. 21 and 121).

Hermann Merxmüller (1920–), a visiting botanist from Germany, collected some 6000 specimens in Namibia. He edited the *Prodromus einer Flora von Südwestafrica*, and

The Mtamvuna Gorge in KwaZulu-Natal, where Olive Hilliard and Bill Burtt discovered *Plectranthus hilliardiae.* In fact the plant had already been found in 1919 along the Mpahlani River by Hans Justus Thode (1859-1932), but his specimen lay unnoticed in a herbarium at Stellenbosch. It was only after Hilliard and Burtt showed their plant to Dr Codd that it was recognized as a new species

contributed text as well. Willie Giess (1910–), a local Namibian, built up the herbarium at Windhoek to which he gave 15 000 specimens, 5000 of which were collected together with other botanists. Merxmüller and Giess discovered a *Plectranthus* species near Rundu in northern Namibia which was named *P. candelabriformis* by Edmund Launert in 1968.

Plectranthus hilliardiae was collected by Olive Hilliard (1925–), a botanist and streptocarpus specialist from KwaZulu-Natal, and Bill (Brian) Burtt (1913–) in the Mtamvuna gorge. Dr Codd named it in her honour.

Jack (John) Repton (1907–) was assistant director of the Pretoria Parks Department and later a lecturer in horticulture at the Pretoria Technikon. He was an active collector with 10 000 specimens to his name, and was the first to collect *P. mutabilis* at Trigaardspoort near Pretoria; this was named by Dr Codd. (The author was one of Jack Repton's students.)

Henri-Alexandra Junod (1863–1934)was a Swiss missionary and naturalist who went to Mozambique in 1889 to become principal of the Evangelical School at Rikatla. In 1899 he established the Evangelist School at Shiluvane in Mpumalanga. He was responsible for the first collection of *P. rubropunctatus* at Nelshoogte Forestry Station.

Inez Verdoorn (1896–) and Amelia Obermeyer (1907–) were botanists from the National Herbarium (BRI) in Pretoria. They collected *P. xerophilus* at Malip's Drift in Mpumalanga.

Bernard de Winter (1924–), a former director of the BRI in Pretoria, and Otto Leistner (1931–), a botanist and editor of various botanical works, were the first to collect *P. unguentarius* (named by Dr Codd in 1974) near Kaoko-Otavi in the Kaokoveld in 1957.

Vincent Wager (1904–), a plant pathologist and author of a book on frogs (*Rana wageri* is named after him), as a boy collected *P. praetermissus* near Port St Johns in 1921 on one of his frog-collecting expeditions, but his specimens were apparently not quite good enough on which to base a new species. Eventually a Mrs Stutterheim brought live plants to Dr Codd who named it in 1979.

Ernst van Jaarsveld (1953–)has been horticulturist at Kirstenbosch Botanical Gardens since 1976. He collected *P. ernstii* (together with his wife Erma) in Oribi Gorge, KwaZulu-Natal in 1977; *P. malvinus* (together with Alison Bingham) as well as *P. reflexus* (together with Ben Dekker and Elaine Campher) at Port St Johns. He also discovered *P. dolomiticus* in 1983 (named by Codd in 1984). *P. mzimvubuensis* (which he named in 2000) was collected by him and colleagues Phakhamani Xaba, Adam Harrower and Godfrey Zwide on a botanical expedition down the Mzimvubu River in 2003. He has tried to popularize the genus and has also named good wild forms.

Hugh B. Nicholson (1906–), a naturalist in southern KwaZulu-Natal, first discovered *P. oribiensis* (named by Dr Codd in 1980) in the Oribi Gorge.

Fanie Venter (1953–), a former botanist at the University of the North now settled in New Zealand collected *P. venteri* (named by Van Jaarsveld and Hankey in 1997) on norite koppies in Sekukuniland, Limpopo Province.

The twenty-first century

The first five years of the present century mark a starting point for plectranthus breeders with the release of 'Mona Lavender' and the 'Cape Angel' series. Three new species have also been named to date. The first *Plectranthus* named this century was *P. mzimvubuensis* which the author collected in 2000.

David Styles (1968–), is the editor of the amateur botanical journal *Plant Life* and chairman of the Krantzkloof Flora Project, established in 2000 to study the flora of the Krantzkloof Nature Reserve and adjoining areas near Durban. Since then he has collected almost 3000 herbarium specimens and found a number of new species and records close to Durban. He collected a plant in the Msikaba River which Trevor Edwards named *P. stylesii* in his honour.

Trevor Edwards (1960–) is Professor of Botany at the University of KwaZulu-Natal. He was responsible for naming *P. stylesii* and co-authoring a number of other *Plectranthus* species with the present author.

Pieter Winter (1964–) a former botanist from the University of the North and presently at the National Herbarium, SANBI, collected a new *Plectranthus* species from the norite hills in Sekukuniland related to *P. xerophilus* and *P. unguentarius* which he and the author have named *P. porcatus*.

Searching for *Plectranthus* 2

My travels in search of plectranthus began in 1974 with day expeditions from the Lowveld Botanical Gardens in Nelspruit, where I worked as a student horticulturist. In those days I was passionate about indigenous succulents, trees and shrubs – and reptiles. This garden was in the heart of *Plectranthus* and reptile territory, and it wasn't long before plectranthus began to distract me from reptiles. The late Koos Nel, who was curator, and the botanist Elize Buitendach first fired my interest, and Jo Onderstall from Ehlatini, an enthusiastic naturalist, further encouraged me. From 1977 onwards I went on a number of very fruitful longer expeditions, including an extensive journey to the main sites in what was then the Transvaal and Natal; a series of expeditions to the Transkei, an area extremely rich in plectranthus which yielded many new species; and lastly no fewer than seven trips to the Kaokoveld in northern Namibia in search of a single plant.

Collecting for the Lowveld National Botanical Gardens

My first collections were made on 29 October 1974 on a day trip to cool escarpment forest around Sabie and Pilgrims Rest. At Kappiekoppie, a dolomite ridge

While searching for *Plectranthus oertendahlii* in the botanically rich Oribi Gorge in 1977, I came across this interesting plant with succulent stems which was not in flower. It turned out to be a new species, and was named *Plectranthus ernstii* by Dr Leslie Codd

Oribi Gorge is famous not only for its scenic beauty but also for its rich and diverse plant life. Here is found a great concentration of *Plectranthus*, including *P. ambiguus, P. ciliatus, P. ecklonii, P. ernstii, P. fruticosus, P. laxiflorus, P. madagascariensis, P. oertendahlii, P. oribiensis, P. petiolaris, P. spicatus, P. verticillatus* and *P. zuluensis*

and cliff face, there were beautiful *Aloe alooides* and indigenous forest. Here I found *Plectranthus grandidentatus*, a trailing aromatic herb growing among rocks, and this I introduced to the gardens (LNBT 190/74, EvJ50). Later that day on the Vaalhoek road towards Kaspers Nek and in dry bushveld among *Aloe castanea* I collected my second species, the succulent *P. cylindraceus* (204/74 EvJ65). My third species was *P. fruticosus* which I collected on another day trip to Brondal on 22 November 1974 (221/74, EvJ125), and after that I could not stop until I had managed to introduce most of the Mpumalanga *Plectranthus* to the gardens.

On Mariepskop peak I found *P. rubropunctatus* and I also visited many desolate parts in the Barberton and Kaapmuiden districts. My first two new records for the region were *P. purpuratus* subsp. *montanus* from the Barberton Mountains, until then known only from Swaziland, and *P. petiolaris* from Crocodile Gorge, recorded before only from the KwaZulu-Natal coast. These collections aroused much interest and made me realize how unexplored the area still was.

Collecting for Kirstenbosch

During May 1976 I moved to Kirstenbosch to take up the post of curator of succulents, which I still hold. My heart was still with the Lowveld and its plants however, especially *Plectranthus*, and John Winter, the curator at Kirstenbosch, encouraged me to go on a collecting trip. Kirstenbosch was then relatively poor in them, with only *P. ecklonii, P. fruticosus, P. ciliatus and P. hadiensis* var. *tomentosus* actually growing, but there was plenty of suitable shade, and so a trip was planned.

Expedition to Gauteng, Limpopo Province, Mpumalanga and KwaZulu-Natal, February-April 1977

On 27 February 1977 I set out on my first major plant-hunting expedition, the aim being to collect plectranthus and other plants for Kirstenbosch. My wife Erma (then pregnant with our second child) accompanied me in our Datsun 1200 bakkie. Before I left I listed all the plectranthus we did not have and planned the expedition accordingly. The prospect of seeing species such as *P. oertendahlii, P. swynnertonii, P. saccatus* subsp. *longitubus* and *P. aliciae* for the first time filled me with excitement.

We started by collecting in the Matjiesfontein area, and then drove north to Winburg to stay with Oom Mias and Theo de Klerk on their farm Candy, collecting *Euphorbia pulvinata* on Tabaksberg. After a stop in Johannesburg with Erma's parents, we travelled to Tzaneen in Limpopo Province where we found *P. swynnertonii* and *P. dolichopodus* at the Debengeni Falls on 10 March. These Falls are in an area of afromontane forest on the Drakensberg escarpment where the rainfall is high. We did not have to look long before spotting *P. swynnertonii* growing among *Impatiens sylvicola* on the forest floor, and we found it again in the moist afromontane forest at Entabeni on the Zoutpansberg. We then made our way through dry bushveld to Wyliespoort and Waterpoort. Here we saw the small-leaved form of *Plectranthus cylindraceus* and *P. tetensis*, and also the stunning red-flowered *Ruspolia hypocrateriformis* and many *Commiphora* species along the northern sandstone slopes.

Our next stop was the Kruger National Park where we had been granted permission to collect the beautiful *Dombeya kirkii*, a large shrub only known (within our borders) from Pafuri which I wanted to introduce to Kirstenbosch. On our way from Schingwedzi we went though a drift which holed our radiator, but luckily we saw a game ranger's lapa on the other side of the stream. We subsequently learnt that the section of river we crossed was the haunt of a large crocodile! After collecting all our target species we left for Echo Caves where we found *P. xerophilus*, then on to Kowyns Pass for *P. rubropunctatus*.

We now made for the Lebombos, discovering en route the Lebombo form of *P. verticillatus* in dry savannah near Mkuze, before turning northwards to Ingwavuma on the Swazi border. Our aim was to locate *P. saccatus* var. *longitubus* which was recorded by Denzil Edwards (1929–) growing in Gwaloweni Forest on top of the Lebombos. This part of the mountain consists of damp subtropical forest which gets moist air and fog from the Indian ocean. The small path was very rough but our bakkie made it, and we were rewarded by the sight of the plant in flower with its beautiful long-tubed corolla and large soft leaves.

It was fairly common in the shade of the forest.

Our next target was *P. psammophilus* from the dry sand forest on the Makatini Flats. This species is perhaps the least gardenworthy, with its small inconspicuous flowers which are prone to rust in cultivation. After collecting our plants we visited Tony and Elsa Pooley at their crocodile breeding farm at St Lucia, where Elsa was to show us the forested areas. On the dunes we found *P. purpuratus* subsp. *tongaensis*, and after a narrow escape from a hippo in the forest we went south to Twin Streams where Ian Garland gave us cuttings of his white-flowered *P. saccatus* subsp. *saccatus* 'Mtunzini'.

Our next and perhaps most important destination was the Oribi Gorge in southern KwaZulu-Natal, an area rich in plectranthus and containing some of the most beautiful. Here we hoped to find *Plectranthus oertendahlii* which had long been in cultivation in Europe but not in South Africa. Autumn was closing in, the date was already the 5th of April, and we did not have much time left. In the shrub forest on top of the sandstone escarpment we searched unsuccessfully among the flat rocks and boulders, but down at the river we found *P. saccatus* subsp. *pondoensis* trailing from the cliffs, and many others such as *P. fruticosus* and *P. verticillatus*. On the margin of the escarpment, growing among *Aeollanthus parvifolius*, we also found a small plectranthus with a potato-like tuberous grey base. At first I thought this was an odd form of the *Aeollanthus* often found in these habitats, but the plant was not in flower. In fact this little cliff-dweller turned out to be a new species which Dr Leslie Codd named *Plectranthus ernstii*. It has since become a popular house plant in Europe and Japan, and is today commonly called the bonsai mint.

In the meantime I resumed the hunt for *P. oertendahlii*. While Erma drove, I jogged down the road leading to the gorge, examining every possible site, until at last I found it with its beautiful variegated rounded fleshy leaves. We continued to the bottom of the gorge collecting *P. fruticosus*, *P. ecklonii* and *P. saccatus* subsp. *saccatus*, but we also came across another plectranthus resembling *P. fruticosus*, bearing peculiar heart-shaped leaves and hairy mauve flowers. When I took cuttings I noticed its unusual hollow internodes. This turned out to be another new species, which Dr Codd named *P. oribiensis*. I discovered later that it had already been collected by Hugh Nicholson who had it in his herbarium.

This was one of my most productive expeditions – in the five weeks I had made 493 live collections.

On a summer trip like this, plants tend to rot rapidly, so we posted material back to Kirstenbosch where my colleagues nursed the cuttings until my return. When I got back I read widely but could find nothing in the literature that matched my cliff-dweller. It was exciting to watch the cuttings turn into plants, and when the stems started swelling and the unique mauve flowers appeared, I was sure that I had a new species. I sent specimens of this and the other plant with hollow internodes to Dr Codd, and he confirmed that both species were new.

First Port St Johns and Mtamvuna expedition, April 1989

This turned out to be a very fruitful two-week expedition which led to the discovery of *Plectranthus reflexus* at Port St Johns, *Gasteria ellaphieae* at Kouga Dam, and southernmost records of *P. ernstii* and *Gasteria croucheri* at Msikaba. I was joined by Elaine Campher and our aim was to study *Gasteria* (which we were revising), endemic trees and shrubs, and *Plectranthus*.

After visiting Robberg at Plettenberg Bay, we headed for Cape St Francis to hunt for *Rapanea gilgiana*, a rare endemic shrub that we wanted to introduce to Kirstenbosch. After this we drove to Kouga Dam where I hoped to locate *Aloe pictifolia* which had been recorded from this site. The vegetation is typical Eastern Cape succulent thicket containing *Portulacaria afra*, *Aloe striata*, *Boschia oleoides*, *Smelophyllum capensis* and two local endemics, *Cussonia gamtoosensis* and *Sterculia alexandri*. The area was very rocky with cliffs, and as I climbed the steep western slope above the parking lot I searched in vain for the little spotted aloe. Sitting underneath a *Sterculia* I saw something to make me start: a small *Gasteria* with beautifully marked leaves which I knew at once was new. I later named this plant *G. ellaphieae*.

After a brief visit to Sapkamma for *Crassula arborescens* subsp. *undulatifolia* we proceeded, via Port St Johns where Ben Dekker joined us, to Mkambati where we collected fruit of the rare Pondo palm (*Jubaeopsis cafra*) on the banks of the Msikaba and Mtentu rivers. This was a wonderful place, sour coastal grassland with sandstone outcrops and many fynbos relics such as *Protea*, *Leucadendron*, *Erica* and *Leucospermum*. We canoed to the palms and I used ropes to climb to their canopies and collect seed. We then went to Forster's Foley (Mtamvuna) where Tony Abbott, who lived in the area, had found *Plectranthus ernstii*. Because of its great horticultural potential I was hoping to find different forms. Although those we found with Tony's help had darker flowers, they were not as fine as the original collections from the Oribi Gorge which had more compact growth, very succulent stems, and a more intense fragrance.

Back in Port St Johns Ben Dekker whetted our appetites with talk of a new lily he'd seen at the Bulolwe River close to his beach house, in an area

1

2

3

1 Mount Thesiger with the mouth of the Mzimvubu River below, near Port St Johns. This is another extremely rich *Plectranthus* area, with eleven recorded species

2 Pugs Xaba and Adam Harrower negotiating rapids in the Mzimvubu River (Ndini region) in 2002. Rubber canoe is often the easiest means of travel in these parts. This expedition yielded a new *Plectranthus* species, later named *P. mzimvubuensis* after the river, as well as two other new species and a white-flowered *Strelitzia reginae*

3 A waterfall near the confluence of the Mzimvubu and Mzintlava Rivers. The local Pondo people warned us of its presence just in time, and helped us find a safe route past it

densely overgrown with evergreen forest trees. Early the next morning we set off upstream, and were delighted to see the rare endemic *Impatiens flanaganiae*, a deciduous tuberous species. We were now in the heart of plectranthus country, and I was amazed at the diversity. Here were *P. zuluensis*, *P. fruticosus*, *P. ecklonii*, *P. praetermissus*, *P. ambiguus*, *P. ciliatus* and *P. saccatus*. Along the path I noticed another erect shrubby plectranthus, and this time I knew straight away that it was new. It had tuberous roots, smooth purplish stems and leaves, and distinctly reflexed floral lobes, so when Trevor Edwards and I later named it we called it *P. reflexus*. Although Ben's 'lily' turned out to be *Haemanthus albiflos*, the expedition was declared a great success.

Second Port St Johns and Msikaba expedition, July 1989

I returned to the area in July because I wanted to find out whether *Gasteria* occurred anywhere in the vicinity of the Mtentu and Mskaba Rivers, Mount Sullivan and Port St Johns. I also hoped to discover more sites of *Plectranthus reflexus*. I was accompanied by Alison Bingham who was horticulturist at the University of Natal Botanical Garden at the time. From Bulolwe we climbed Mount Sullivan from the south-west, and as we were traversing the steep upper slopes and cliffs near the peak, Alison pointed out an interesting, rough, succulent-leaved *Plectranthus*. At first I thought this was a form of *P. ciliatus*, and called it 'Bingham' after her, but Trevor Edwards and I later decided that it was a species in its own right, and we named it *P. malvinus* because of its bright mauve flowers.

After Mount Sullivan we went to the Mtentu River where we found *Gasteria croucheri*. In the afternoon we rode on horseback to inspect the Msikaba Gorge, which was overgrown with thicket. The cliffs, however, were rich in succulent vegetation, and we stopped at several sites where we could gain access. I was fascinated to find *Plectranthus ernstii* here in a more slender form on rocky ledges growing with *Gasteria croucheri*, the yellow *Delosperma edwardsiae*, *Senecio medley-woodii*, *Cotyledon velutina* and *Plectranthus saccatus* subsp. *pondoensis*. This site marks the southernmost limit of the *Gasteria* and *P. ernstii*.

The Mzimvubu River expedition, 2002

The Mzimvubu, with its mouth at Port St Johns, is the largest river system in the Transkei and part of the *Plectranthus* centre of endemism. It also contains many cliffs, and at the time I was studying cliff-adapted succulents. The deep river systems in the Transkei have yielded many new plants in the past, and I knew that the inner parts of the Mzimvubu were inaccessible by motor car, so the only way was to travel by canoe.

I gathered together a team consisting of our yearly Kirstenbosch scholar, Phakhamani Xaba, my colleague Adam Harrower and Godfrey Zwide from Eastern Cape Nature Conservation.

We launched our two rubber canoes at Welch Bridge (about 25 km south of the Umtata–Mount Frere national road) and paddled down the river, stopping at interesting cliffs and camping during the night. Sometimes we had to drag our canoes across obstacles, and we overturned several times, losing equipment, but luckily not plants. The vegetation in the area is mainly grassland and dry kloof with forest and thicket at times. The local Pondo were very curious about us, and helped us negotiate the large waterfall lower down. We collected many interesting plants (including a southernmost record for *Gasteria croucheri*, and a northernmost for *Crassula foveata*), some of which turned out to be new.

On our last day we stopped off at a river bend opposite Oeza below Ludonga at a cliff composed of Ecca shale. As I scrambled up the scree below the cliff, I saw the flat linear leaves of an exciting little *Drimia* with basal club-shaped leaf scales. While I was looking at it, I became aware of a scandent pink-flowered shrub dangling in front my eyes, and when I looked up I saw it was a plectranthus. Here were two new species growing together on a cliff. The drimia received the name *D. mzimvubuensis*, and I named the pink plectranthus *P. mzimvubuensis*, and now grow it at Kirstenbosch. It has turned out to belong to the section *Plectranthus* and is the first with rounded subterranean potato-like root tubers produced away from the stem towards the end of the roots, as I discovered while growing it.

The Kaokoveld search for *P. unguentarius*

By the late eighties I had collected all the known South African plectranthus, but there remained a Namibian one that had always intrigued me. *P. unguentarius* was discovered by Bernard de Winter and Otto Leistner south of Kaoko-Otavi in 1957. From their description it was an erect, pleasantly aromatic, conspicuous shrub with white flowers which would be easy to find. I planned to go into the area, see the plants, collect material and come back. In fact it took seven expeditions to find this plant, but my searches bore fruit in the end, and on nearly all expeditions I was rewarded with a new record or a new species.

My first expedition was in April 1990 with Ben Engelbrecht, then curator of the Karoo National Botanical Gardens, and Chuck Hanson, a visiting American from Tucson well known for his success with *Euphorbia* cultivation, who sponsored our expedition. We went to many sites in search of this elusive species. We went as far as the arid mopane savannah of Sesfontein and inspected the Khowarib Poort, which consists of dolomite cliffs, sure we would find it there. We did find *Aloe dewinteri* on the cliffs, *Ceraria longipedunculata* and *C. carisoana*, however, which was some consolation.

For the next seven years I was too busy with my *Gasteria* book and other writing, as well as with planting up the new Conservatory at Kirstenbosch to return to Namibia. However once things quietened down I found I still needed to get Namibian succulents for the Conservatory. An expedition was arranged for mid-winter, and this time I thought I would make sure of *Plectranthus unguentarius* by going straight to the location given by de Winter and Leistner, 17 km south of Kaoko-Otavi. But the roads had changed since then, and although everything looked right, we were to be disappointed. The dry mopane woodland among dolomite rocks was filled with many interesting rare succulents, *Aloe dinteri* in particular, but not our plant, and we had to leave.

Three years later I began a study of cliff-face plants and decided to visit the cliffs in Kaokoveld as part of my research. An expedition was planned for July 2000, and my companions were Trevor Adams from SANBI, Freddie van Wyk, Anton Cilliers and Philip Griffiths, a horticulturist from Kew. This time I was convinced I would find the plectranthus in the Otjihipa Mountains among dolomite fragments where (according to Dr Codd) Davies, Thompson and Miller had collected their plants. We searched in the vicinity of Epupa, then took the Otjiboronbongo road to the easily found site of *Aloe corallina*, one of my target species, which was plentiful on the black dolomite cliffs. The area is very mountainous, the road atrocious, and we nearly did not make it back. That done, we resumed our search for our lost plectranthus. After climbing a few peaks in vain, we made a last try for Koakora Spring, a promising area. The vegetation is arid savannah and very mountainous. We aimed for the highest peak, but on the black dolomite cliffs in the kloof I soon spotted a large plant belonging to the Lamiaceae with square succulent stems and large leaves. This had to be our plant and we celebrated – too soon, as it turned out. When we searched for dry floral fragments it became clear that the plant in question was a tetradenia, although a new summer-flowering one, which Braam van Wyk and I later named *Tetradenia kaokoensis*. We returned to Kaoko-Otavi, thinking we must have overlooked the plant there, but in vain.

I repeated the expedition in 2001, and found a number of other plants, but not this one.

On our 2002 expedition, we discovered Omavanda (Baynes Mountains) with its plant treasures. Here was a new aloe, *Aloe omavandae*, as well as *Aeollanthus rydingianus*, *Plectranthus dinteri* on the cliffs and *P. hereroensis* in the

1

2

3

1 The author collecting plants on a cliff-face at Omavanda, in the south-eastern Baynes Mountains in the Kaokoveld. The search for the elusive *Plectranthus unguentarius* involved many expeditions to this area. On this occasion the party found a new species of aloe, *A. omavandae*

2 Camping in Slangpoort in the central Baynes Mountains in 2005

3 The Slangpoort gorge. This narrow and dangerous section is inhabited by a number of snakes, including the dwarf python *(Python anchietae)*, and is a death trap for animals such as the Kaoko dassie. We covered the 15 km of Slangpoort on foot, finding *Plectranthus hereroensis* as well as many other new plants, some still as yet unidentified

kloofs. We then decided to go down Van Sijl's Pass to the western Otjihipa mountains into a very arid zone just before the true Namib. On entering a kloof east of the Ezorotuuo, we came across some Otjihimba people to whom we showed a picture of the plant. They said they knew it, but wanted money. After about 300 m they showed us an aromatic *Petalidium*, and I realized they did *not* know our plant. The location in any case was granite, and I was sure the plant grew on dolomite.

We continued to walk upstream, finding water plentiful after a good rainy season. Our aim was a high peak about 15 km to the east, and although we did not get there, we did find another large new aloe (which we later named *Aloe kaokoensis*) on the steep slopes, and a peculiar cliff-dwelling herb which turned out to be *Rogeria petrophila*, originally discovered by de Winter and Leistner.

The next morning we met Koos Verwey at his Syncro Tourist Camp, where he helps the Himba people sell their craftwork to tourists to fund a school. He called all the older medicine people together and when they saw a picture of a dried specimen of *P. unguentarius*, they called it Omumbumbwa. We were very excited as we now had a name. They suggested we try again at Kaoko-Otavi and Opuwo. After an exhausting journey to both places we discovered their name referred to a type of *Commiphora*, the bark of which they use!

In 2003 we climbed the Brandberg for *Plumbago wissii* and other cliff-dwelling plants.

In 2004 I planned my sixth expedition, this time with Wessel Swanepoel from Windhoek. I was curious to see an unnamed *Pelargonium* collected by the Pretoria botanist Pieter Winter on L. G. Thompson's 1996 Kaokoveld expedition. After camping in the Omavanda cave and climbing the highest peak in Kaokoveld, we again took the Van Sijl's Pass road to Koos Verwey's camp, this time heading for the peak of Otjihipa. Otjihipa is a large steep twin-peaked mountain just under 2000 m, with its base about 300 m above sea level. The climate is very hot and the vegetation is desert, so in the dry season it is essential to carry water. Because of the distances we decided to do it in two to three days carrying at least 10 litres of water each. We climbed the western peak, which is slightly lower than the eastern peak, but found no plectranthus although we did come across a new cliff-growing *Aeollanthus*. We also collected a shrubby *Othonna* (a fynbos relative) on the cliffs as well as a succulent-stemmed *Cynanchum*, and I was glad to rediscover Pieter's *Pelargonium*. By now the heat was intense, our water was running low, and we had to abort the trip.

The *Aeollanthus* was treasured, propagated, and then accidentally thrown away at Kirstenbosch.

We decided to plan a seventh trip for January 2005, the rainy season, to recollect the *Aeollanthus* and also find a *Dalbergia* species – we had entirely given up on the plectranthus. Joining us were Wessel Swanepoel, Barbara and Jan Burring. This time Braam van Wyk sponsored the expedition and we persuaded him to join us. Using our cave at Omavanda as a base, we made several collections including *Plectranthus dinteri* on the granite cliffs as well as the new *Aeollanthus*, and *Tetradenia kaokoensis* on the sheer east face. This was quite a surprise as we'd thought it was a dolomite endemic. It rained continually for two days, putting my digital camera out of action. After Omavanda we again travelled towards Van Sijl's Pass, passing the dry Omuhonga river bed and a huge ana tree, *Faidherbia albida*. As we saw the first granite outcrop I spotted *Plectranthus unguentarius* out of the corner of my eye. I shouted 'Stop!', ran out and there they were, on a granite outcrop where I had least expected them. They were unmistakeable plants after all, large erect shrubs, in full leaf and still with the remnants of old dry inflorescences towering three metres in the air, the succulent leaves with the most rewarding fragrance.

The next day Wessel and I climbed Otjihipa, often in rain, and at times we had to wait for the fog to lift. Water was plentiful and we were much more mobile with less weight on our backs. This time we again found our new *Aeollanthus* and also *Plectranthus dinteri* and *Tetradenia kaokoensis* on the granite cliff-faces just below the peak. We even managed to climb both peaks. On our return to base camp, Koos Verwey warned us that the Hoariseb River was in flood, so we had to change our route back, but this enabled us to see two *Crinum* species in flower and the endemic *Turnera ocellata* as well.

Back at Kirstenbosch our cuttings of *P. unguentarius* rooted rapidly. By April it was in flower and Vicki Thomas was able to illustrate it.

***Plectranthus unguentarius*, found at last in 2005 after a ten-year search. It was growing on a granite outcrop just east of the inland escarpment, in northern Kaokoveld, Namibia. It was plentiful, but highly localized**

3 Plectranthus in the garden

Opposite: *Plectranthus ambiguus* 'Oribi' growing under a klip-kershout (*Maytenus oleoides*)

***Plectranthus neochilus* 'Pietersburg' is a useful groundcover for sun**

Most of the South African plectranthus make first-rate garden and house plants. They are all summer growers, starting their cycle during spring, flowering in late summer or autumn when the daylength is shrinking, and inactive in winter when they can pruned back. They grow rapidly, respond to fertilizers, and apart from eelworm are relatively pest-free. Although they do not like frost, they can survive light frost under trees, and even if they are damaged will simply resprout during spring. It is remarkable that of the 350 species in the world, the most gardenworthy of these are local, and the choicest are from the Eastern Cape and the adjacent southern KwaZulu-Natal.

As far as gardeners are concerned, plectranthus divide into four groups. There are the shrubs which generally prefer shade, with a few interesting exceptions; the groundcovers which also prefer shade though again

***Plectranthus ecklonii* 'Erma' (pink), 'Medley Wood' (blue), and 'Tommy' (white)in the main garden at Kirstenbosch. The three forms make a colourful combination. They are all easy, fast-growing plants**

Opposite: Various colour forms of *Plectranthus saccatus* grow together happily in the Stellenbosch garden of the *Plectranthus* breeder Gert Britz

there are exceptions; the succulents for rockeries; and another group that makes good house plants. In the wild the largest group comes from the shady moist coastal subtropical forest, the next largest from the afrotemperate regions, and the rest are either bushveld species or sunny grassland, the smallest group.

When deciding which to grow, it's always best to consider one's local climatic conditions. If you live in a cooler moist coastal climate, the species suitable for the forest garden would be ideal. If yours is a dry bushveld climate, it is best to look for the bushveld species. If you live on the highveld you have a wider choice as long as you provide some form of frost protection.

The forest or subtropical garden

If you live in KwaZulu-Natal or on the Mpumalanga escarpment, the forest plectranthus are an ideal choice for an afromontane or forest garden. They also thrive in the cooler moister parts of the Western Cape, and along the Garden Route. They grow well in Port Elizabeth provided they are given summer water and extra feeding. Some are fairly robust and can be grown on the highveld. *P. neochilus* is the toughest, and thrives throughout South Africa, tolerating full sun.

Even in a sunny garden one can find a suitable microclimate for these plants, such as the south side of a wall where it is cooler with less sun. The forest species generally require slightly acid soil. They thrive on compost and natural non-burning organic products such as Neutrog, which we use with great success at Kirstenbosch.

Shrubs

The recommended shrubby species for the forest garden are *Plectranthus fruticosus*, *P. ecklonii*, *P. zuluensis*, *P. saccatus* and *P. hadiensis* var. *tomentosus* 'Carnegie', a sun-tolerant species which is decumbent, meaning that the stems lie along the ground before turning upwards. These are tall plants, best placed in the background with the smaller species towards the front. Young plants set in the ground during late winter or spring will flower the first autumn.

The three cultivars of *P. ecklonii* 'Medley Wood' (blue), 'Erma' (pink) and 'Tommy' (white) are stunning when grown in large groups. *P. fruticosus* also has superb forms. 'Behr's Pride' is a great hit, with pink flowers and leaves with beautiful dark purple undersides. The stems will eventually grow up to 2 m if left unpruned, but they are best cut hard back straight after flowering. 'Ellaphie' has decumbent growth (it trails with the tips upright), so can be grown as a groundcover, while 'James' will tolerate full sun. Its leaves have a frilly margin and beautiful purplish veins. *P. saccatus* makes an unpruned shrub to 2 m; subsp. *pondoensis* has succulent leaves and small flowers. Its form 'Myokane White' has white flowers faintly spotted with mauve and groundcovering spreading growth. Of the two long-tubed forms of *P. saccatus* subsp. *saccatus*, 'King Goodwill' has shiny leaves and white flowers, while 'Ngoye' is similar but with mauve flowers. *P. saccatus* subsp. *saccatus* 'Nkandla' has a hooded upper lip. *P. zuluensis* is also a rewarding species; the best form is 'Oribi Gorge' with attractive flowers, while 'Sky' has dark blue flowers. With pruning, and in certain areas, these can be grown as groundcovers. Suitable companion plants are *Clivia miniata*, *Chlorophytum comosum*, *Dracaena aletriformis*, *Scadoxus multiflorus*, *S. membranaceus*, and the bosplakkie, *Crassula multicava*; there are more suggestions in the next chapter.

Plectranthus shrubs and perennials for shade

Name and cultivar	Height	Growth rate	Flower	Fl. time	Leaves	Cold tolerance
P. barbatus	1–3 m	medium	blue-purple	Dec–April	green	mild frost
P. barbatus 'Vicki'	1–3 m	medium	blue-purple	Dec–April	variegated	mild frost
P. dolichopodus	25–70 cm	fast	sky-blue	Oct–May, peak in March	green	mild frost
P. ecklonii 'Erma'	1–1.5 m	fast	pink	March	green	mild frost
P. ecklonii 'Medley Wood'	1.5–2 m	fast	blue	March	green	mild frost
P. ecklonii 'Tommy'	1–1.5 m	fast	white	March	green	mild frost
P. fruticosus 'Behr's Pride'	1–1.5 m	fast	pink	March	purple below	sensitive
P. fruticosus 'James'	1–1.5 m	fast	pink	March	purple veins	mild frost
P. fruticosus 'Ngoye'	1.5 m	fast	pink	March	rounded	sensitive
P. fruticosus 'Pauline'	1–1.5 m	fast	mauve	March	purple below	sensitive
P. grallatus	0.5–1 m	fast	white	March	hairy	frost-hardy
P. hadiensis var. tomentosus 'Carnegie'	100 cm	fast	light purple	March	hairy	mild frost
P. mzimvubuensis	1–2 m	medium	mauve	March	green	sensitive
P. oribiensis	1 m	medium	purple	April	heart-shaped	sensitive
P. reflexus	1–1.5 m	slow	purplish	Jan–March, peak in March	purplish	sensitive
P. rehmannii	1–1.5 m	medium	white	March–May	green, matt	mild frost
P. rubropunctatus 'Entabeni'	1.5 m	fast	white to light mauve	March	matt	mild frost
P. saccatus subsp. *saccatus* 'Gwaloweni'	1 m	medium	mauve, tube 20–26 mm	Nov–June, peak in March	soft, thin	sensitive
P. saccatus subsp. *saccatus* 'King Goodwill'	1.5 m	fast	white, long-tubed	March	shiny, thin	sensitive

Name and cultivar	Height	Growth rate	Flower	Fl. time	Leaves	Cold tolerance
P. saccatus subsp. *saccatus* 'Kirsten'	1.8 m	fast	purplish-blue	Nov–June, peak in March	matt	sensitive
P. saccatus subsp. *saccatus* 'Mtamvuna'	30–50 cm	fast	mauve, purple speckled, densely flowered	peak in March	matt	sensitive
P. saccatus subsp. *saccatus* 'Mtunzini'	0.5–1 m	medium	white	Nov–June, peak in March	matt	sensitive
P. saccatus subsp. *saccatus* 'Ngoye'	1.5 m	fast	mauve, long-tubed	March	shiny, thin	sensitive
P. saccatus subsp. *saccatus* 'Nkandla'	1 m	fast	mauve, hooded	Nov–June, peak in March	matt, thin	sensitive
P. zuluensis 'Oribi Gorge'	1.5 m	medium	light purple-blue	Oct–June, peak in March	matt	sensitive
P. zuluensis 'Sky'	0.5–1 m	medium	dark purple-blue	Oct–June, peak, in March	matt	sensitive
P. zuluensis 'Umgai'	2 m	medium	dark purplish-blue	Oct–June, peak in March	matt	sensitive

1

Groundcovers

There are many groundcovering species suitable for shade. One of the most attractive is *Plectranthus ciliatus* 'Drège' with its dark purple undersides. It will form a complete cover with sprays of dainty white flowers in the autumn. *P. malvinus* is similar, but smaller with more leathery leaves and pink flowers. It is more drought-tolerant. Other recommended species are the pincushion spurflower (*P. ambiguus*), the cress spurflower (*P. swynnertonii*) which has a cress-like scent and sharply dentate leaves and white flowers. The thicket spurflower *P. madagascariensis* is ideal for dry shade, and is one of the most drought-tolerant species. *P. oertendahlii* has large round variegated leaves and often behaves as a weak self-seeding perennial. The prostrate *P. saccatus* subsp. *saccatus* 'Izingolweni' is a groundcover with a heavily speckled corolla. Other recommended groundcovers for drier conditions are the gossip spurflower, *P. verticillatus*, and the Dutch-pipe spurflower, *P. petiolaris*.

2

1 *Plectranthus ciliatus* 'Drège'in full flower in the autumn

2 The variegated *Plectranthus madagascariensis* 'Lynne', a useful drought-resistant groundcover for shade

Plectranthus groundcovers for shade

Name and cultivar	Height (cm)	Growth rate	Flower	Fl. time	Leaves	Container	Hanging baskets	Cold tolerance
Plectranthus aliciae	25	fast	white	March	medium	X		sensitive
P. ambiguus 'Karkloof'	10	fast	mauve	March	small	X	X	mild frost
P. ambiguus 'Nahoon'	30	fast	mauve	March	glabrous	X	X	mild frost
P. ambiguus 'Oribi'	10–30	fast	mauve	April	large	X	X	mild frost
P. ambiguus 'Thesiger'	40–60	fast	mauve	April	green	X		mild frost
P. ambiguus 'Umque'	30	fast	mauve	March	purplish	X		mild frost
P. ciliatus 'Drège'	30	fast	white	March	purple below	X	X	mild frost
P. ciliatus 'Richard'	20	fast	pale pink	March	hairy	X	X	mild frost
P. elegantulus	30	fast	white	March	green	X	X	mild frost
P. fruticosus 'Ellaphie'	60–100	fast	mauve	March	purple below			sensitive
P. grandidentatus 'Hlobane'	60	fast	white	March	medium, hairy	X	X	mild frost
P. grandidentatus 'Liesl'	30	fast	white	March	large	X	X	mild frost
P. hilliardiae 'Magwa'	20	slow	mauve	Feb–June	shiny, purple below	X	X	sensitive
P. laxiflorus	30–60	fast	white	March–June	heart-shaped			sensitive
P. lucidus 'Albert'	15	fast	white	March	shiny	X	X	sensitive
P. lucidus 'Bulolwe'	15	fast	mauve	March	shiny	X	X	sensitive
P. madagascariensis	20	medium	white	March	green	X	X	mild frost
P. madagascariensis 'Lynne'	20	medium	white	March	variegated	X	X	mild frost
P. *malvinus*	20	medium	mauve	March	green, rough	X	X	sensitive
P. oertendahlii 'Mzimkulu'	10	medium	white	Feb–May, peak in April	rounded, variegated	X	X	sensitive
P. oertendahlii 'Uvongo'	10	medium	white	Feb–May, peak in April	large, rounded, silvery green, variegated	X	X	sensitive
P. petiolaris	30	medium	violet-purple	April	green			sensitive
P. praetermissus	20–30	fast	mauve	March	glossy	X	X	sensitive
P. purpuratus subsp. *tongaensis* 'Kosi'	10–20	fast	white	Sept–May, peak in April	dark purple below	X	X	sensitive
P. rubropunctatus 'Mariepskop'	20–30	fast	white	March	matt			mild frost
P. saccatus subsp. *pondoensis*	30–50	medium	mauve, short tube	Nov–April, peak in March	succulent	X	X	mild frost
P. saccatus subsp. *pondoensis* 'Myokane White'	30–50	medium	white, spotted with mauve, short tube	Nov–April, peak in March	succulent	X	X	frost-tolerant
P. saccatus subsp. *saccatus* 'Izingolweni'	20–30	fast	mauve, dense purple speckles	Peak during March	matt	X	X	sensitive
P. strigosus	5	slow	white	Sept–May, peak at end of March	matt	X	X	sensitive

Name and cultivar	Height (cm)	Growth rate	Flower	Fl. time	Leaves	Container	Hanging baskets	Cold tolerance
P. stylesii 'Msikaba Gorge'	12–15	slow	white	Sept–May, peak in April	matt, succulent	X	X	sensitive
P. stylesii 'Mtamvuna'	12–15	slow	white	Sept–May, peak in April	shiny	X	X	sensitive
P. swynnertonii	20–30	fast	white	Sept–May, peak in April	matt	X	X	sensitive
P. swynnertonii 'Woodbush'	20–30	fast	pink	Sept–May, peak in April	matt	X	X	sensitive
P. verticillatus 'Barberton'	10–15	fast	white	Sept–May, peak in April	shiny, dark purplish below	X	X	mild frost
P. verticillatus 'Blyde'	10–20	fast	white	March	large, glossy	X	X	mild frost
P. verticillatus 'Eureka'	10–20	fast	white	March	hairy, purple below	X	X	mild frost
P. verticillatus 'Freckles'	10–20	medium	white	March	variegated, spotted	X	X	mild frost
P. verticillatus 'Gossip'	10–20	medium	white	March	variegated	X	X	mild frost
P. verticillatus 'Malelaan'	20–30	fast	pink	March	large, hairy	X		mild frost
P. verticillatus 'Money Maker'	5–10	fast	white	Dec–April, peak in March	small, round, matt	X	X	mild frost
P. verticillatus 'Pink Surprise'	12–15	fast	pink	Dec–June, peak in March	smooth	X	X	mild frost
P. verticillatus 'Ubombo'	3–10	fast	white	March–April	shiny	X	X	light frost

1 *Plectranthus tetensis* is a vigorous vegetative annual good for dry bushveld gardens

2 *P. verticillatus* 'Ubombo' makes a useful groundcover in dry shady gardens and also does well in hanging baskets

3 *P. oertendahlii* 'Uvongo'. This beautiful cultivar was introduced in April 1979 from Uvongo, southern KwaZulu-Natal

4 *P. ambiguus* does well in a hanging basket

5 *P. ernstii*, the bonsai mint, is easily grown from cuttings, rapidly developing its succulent stems

1

The bushveld garden

There are a fair number of plectranthus suitable for warm bushveld gardens, either in full sun or under trees. It is best to maintain the bushveld character by using flat-topped acacia trees, erect nabooms such as *Euphorbia ingens* or *E. candelabra*, and different aloes, especially *Aloe marlothii*, *A. castanea*, *A. aculeata* and *A. cryptopoda*. Smaller companion plants could include *Aloe vanbalenii*, *A. zebrina*, *Eulophia petersii*, *Sansevieria pearsonii* and *Kleinia fulgens*.

Shrubs

These shrubby species are best grown in full sun. The tall *Plectranthus unguentarius* makes an erect deciduous shrub to 3 m. It has stunning greenish-white flowers and pleasantly aromatic leaves which fall after flowering is over. The purple-flowered Penge spurflower (*P. xerophilus*) and white-flowered Sekukuni spurflower (*P. porcatus*) are otherwise similar; both grow to about 1–2 m. All these three have tuberous roots and are therefore very drought-tolerant. The Herero spurflower (*P. hereroensis*) has tall rounded growth, and is best raised annually from seed sown during spring. It grows to 1.5–2 m. *P. mirabilis* is also a stunning shrubby species to 3 m, with dark blue flowers and pale leaf undersides.

Groundcovers

Suggested groundcovers include *P. verticillatus*, *P. mutabilis*, *P. petiolaris*, *P. neochilus* and *P. tetensis*.

Rock garden plants

The following species can be grown among rocks in coastal areas: *Plectranthus cylindraceus*, *P. ernstii*, *P. neochilus* and *P. spicatus*. Although *P. venteri* is a bushveld species, it thrives in rockeries of the Western Cape and elsewhere.

The best rockery species for bushveld gardens include *P. amboinicus* which has pleasantly aromatic

2

3

leaves, *P. cylindraceus*, *P. purpuratus* (especially its subsp. *purpuratus* and *montanus*, but also the groundcovering subsp. *tongaensis*), *P. spicatus* and *P. venteri*.

Pot plants and hanging baskets

Many plectranthus are suitable for pot plants and hanging baskets, especially the groundcovers or trailers. They can be grown in a standard houseplant mixture which should be rich in organic matter, and they can be fed with an organic fertilizer during spring and summer. They will all need regular watering. It is a good idea to pinch out the tips to maintain a neat shape. Repotting is best done in late winter, when one can inspect the roots for eelworm, especially if the plant is losing vigour.

Recommended species are *Plectranthus ambiguus*, *P. ciliatus, P. elegantulus*, *P. hilliardiae* 'Magwa', *P. malvinus*, *P. oertendahlii*, *P. praetermissus*, *P. purpuratus* subsp. *montanus* and subsp. *tongaensis*, and *P. verticillatus*. The bonsai mint, *P. ernstii*, is also suitable for pots. Pinch out the tips regularly to promote the succulent stems and the 'bonsai' look.

4

5

Plectranthus shrubs and perennials for sun

Name and cultivar	Height	Growth rate	Flower	Fl. time	Leaves	Container	Cold tolerance
P. barbatus	1.5–3 m	fast	violet	summer, peak in March	green	X	mild frost
P. barbatus 'Vicki'	1.5–3 m	fast	blue-purple	summer, peak in March	variegated	X	mild frost
P. candelabriformis	1 m	fast	violet	March	green	X	mild frost
P. fruticosus 'James'	1–1.5 m	fast	pink	March	purplish veins	X	mild frost
P. crassus	60 cm	medium	blue	April-Aug	grey green	X	mild frost
P. hadiensis var. *tomentosus* 'Carnegie'	100 cm	fast	light purple	March	hairy	X	mild frost
P. hereroensis 'Kaoko'	1.5–2 m	fast	dark purple, large	March–May	green	X	mild frost
P. hereroensis 'Malmokkie'	1.7 m	fast	white, small	March–May	green	X	mild frost
P. hereroensis 'Matoppos'	1.7 m	fast	dark purple, small	March–May	green	X	mild frost
P. hereroensis 'Witpoortjie'	1.7 m	fast	light blue	March–May	green	X	mild frost
P. mirabilis	2–3 m	fast	blue	autumn	white below	X	mild frost
P. porcatus	1–2 m	medium	white	autumn	rough	X	mild frost
P. sanguineus	45 m	slow	dull mauve	March	succulent	X	mild frost
P. venteri	30–60 cm	medium	mauve	summer, peak in March	dissected	X	mild frost
P. unguentarius	1–3 m	fast	white to cream	autumn	sweetly aromatic	X	mild frost
P. woodii	30–50 m	fast	white	autumn	rough	X	frost-hardy
P. xerophilus	1–2 m	medium	purple	March–June, peak in April	leathery	X	mild frost

Plectranthus purpuratus subsp. _purpuratus_ is a cliff-dweller with a remarkable ability to adjust its leaf size and arrangement according to the amount of light and moisture it gets. Under stressful exposed conditions the succulent leaves, which are purplish below, become imbricate (densely overlapping), as here

Plectranthus groundcovers for dry sun or partial shade (mainly succulent species)

Name and cultivar	Height (cm)	Growth rate	Flower	Fl. time	Leaves	Container	Hanging baskets	Cold tolerance
P. amboinicus 'White Fragrance'	30	medium	white	March	variegated	X		mild frost
P. cylindraceus 'Lebombo'	30	medium	white	March	densely hairy	X		mild frost
P. cylindraceus 'Wylie'	25	medium	white	March	small, hairy	X		mild frost
P. dinteri	30	fast	light mauve	Dec–April	large teeth	X	X	mild frost
P. dinteri 'Waterberg Maroon'	40	fast	dark maroon	Dec–April	large teeth	X	X	mild frost
P. dolomiticus	20	medium	violet-purple	Dec–May, peak in March	succulent	X	X	mild frost
P. ernstii 'Forster's Foley'	10	slow	blue	Oct–April	light shade	X	X	sensitive
P. ernstii 'Msikaba'	10	slow	white	Oct–April	light shade	X	X	sensitive
P. ernstii 'Sikuba'	10	slow	light mauve, purple stripes	Oct–April	light shade	X	X	sensitive
P. ernstii 'Oribi'	15	medium	light mauve	Oct–April	light shade	X	X	sensitive
P. esculentus	20–30	medium	yellow	Aug–Sept	deciduous in winter			mild frost
P. hadiensis var. *tomentosus* 'Mamba Valley'	30	medium	white	March	very hairy			mild frost
P. hadiensis var. *tomentosus* 'Shongweni'	40	fast	white	March	wedge-shaped	X	X	mild frost
P. madagascariensis	20	medium	white	March	succulent	X	X	mild frost
P. mutabilis	10	fast	purple-blue	Feb–April	purplish	X	X	mild frost
P. neochilus 'Pietersburg'	20	fast	purplish-blue	peak in spring	very succulent	X		mild frost
P. neochilus 'Siteki'	10	medium	light blue	March	succulent	X		mild frost
P. ornatus	20–35	medium	purple-blue	summer, peak in autumn	succulent	X	X	mild frost
P. pentheri	10	medium	light blue	March	succulent	X		mild frost
P. purpuratus subsp. *montanus* 'Lomati'	10	medium	white	Sept–May, peak in April	succulent, hairy	X	X	mild frost
P. purpuratus subsp. *purpuratus* 'Umgeni'	10	medium	white	Sept–May, peak in April	rounded, succulent, purplish	X	X	sensitive
P. saccatus subsp. *pondoensis*	30–50	medium	mauve, short tube	Nov–April, peak in March	succulent	X	X	mild frost
P. saccatus subsp. *pondoensis* 'Myokane White'	30–50	medium	white, spotted with mauve, short tube	Nov–April, peak in March	succulent	X	X	frost-tolerant
P. spicatus 'Nelspruit'	20–30	slow	violet-purple	March–June, peak in April	succulent	X		sensitive
P. spicatus 'Ubombo Mountain'	20–30	slow	violet-purple	March–June, peak in April	succulent	X		mild frost
P. tetensis	5	fast	purple	March to May, peak mid-April	succulent	X	X	mild frost

1

1 *Plectranthus fruticosus* 'Ellaphie' makes a colourful shrub during autumn

2 *Plectranthus fruticosus* 'Behr's Pride' is one of the most beautiful forms of this plant. It originates from near Lusikisiki in the Eastern Cape

2

Pruning

Although plectranthus will grow and flower without being pruned, it is advisable to prune to keep the growth neat, maintain vigour and maximize flowering. Pruning is a renewal process and occurs in the natural habitat as well. The best time to do it is during midwinter, after flowering and fruiting.

During autumn the inflorescence, which is usually borne in a terminal raceme, spike or panicle, will die back and the next season's growth will sprout from buds that emerge from the nodes below. Using a pair of sharp, clean secateurs, cut back to just above a node. There are no fixed rules and it is hard to go wrong. Even if a plant is cut to the ground it will usually still resprout; this happens frequently in the natural habitat when fires sweep through the undergrowth, and plants generally recover fully. All dead branches and unhealthy growth should also be removed. The softly herbaceous wood is very easy to cut, though one may need a handsaw to remove the older wood at the base. Plectranthus are rapid growers and will resprout very quickly when the weather warms up.

Shrubby species in the forest garden

The tall shrubby species such as *P. ecklonii*, *P. fruticosus* and *P. rubropunctatus* are best cut hard back to a basic framework (roughly above or below knee height), leaving a few nodes from where the new growth will sprout. *P. zuluensis* can be pruned less severely; it can be left to form a dense shrub up to 2 m, and then trimmed according to need. *P. barbatus* can also be pruned according to need. Some forms of *P. saccatus* when left unpruned will make a dense rounded shrubbery of 1,5 m or more high. They can then be pruned to the required height.

Decumbent and leaning shrubs

These include *P. hadiensis*, *P. laxiflorus* and *P. woodii*. They can all be pruned back hard, as in nature they annually succumb to veld fires. The larger robust forms of *P. ambiguus* can also be pruned back hard. The coastal *P. petiolaris* and the thicket *P. madagascariensis* can also be pruned according to need, or they can be left alone and only the dead branches removed. Leave enough buds to sprout.

Groundcovers

P. ciliatus, *P. verticillatus*, *P. swynnertonnii*, *P. grandidentatus*, *P. ornatus*, *P. lucidus*, *P. malvinus* and *P. strigosus* if left will all maintain their cover. Some gardeners prefer to cut these back with a weedeater. They are all vigorous and should recover fast during spring.

Rock garden species

The succulent species such as *P. cylindraceus*, *P. amboinicus*, *P. spicatus* and *P. neochilus* (which is also used as a groundcover) are slower-growing and only need to be trimmed very lightly, chiefly to maintain a neat shape. The tall shrub *P. xerophilus* can be pruned after flowering to the required height. *P. unguentarius* can also be cut back hard.

Pot plants and hanging baskets

Vigorous plants such as *P. hilliardiae*, *P. oertendahlii*, *P. malvinus* and *P. elegantulus* use a lot of energy and need to be annually renewed. To maintain floriferous specimens they should being pruned back hard after flowering and replanted in refreshed soil. *P. ernstii* can be pruned to maintain a 'bonsai' look. To do this the stems tips should be regularly cut back to keep the plant small and compact, although during the autumn it can be left alone to flower.

Propagation

Plectranthus are the easiest plants to propagate. They are adapted to cope with disturbances on the forest floor and even dropped branches will root. Large areas can also be established with unrooted stems in the same way one plants a lawn; with watering, one should get at least 80% rooting. They are also easily raised from seed, and the offspring should show enough variability to enable one to make a good selection.

Seed

About 2–4 seeds are produced in each small, often shiny nutlet, which is usually persistent, so the whole seedhead can be picked. The seeds are easily extracted by rubbing between the fingers. They are relatively long-lived and store well in dry labelled containers.

Sow the seed during spring in a well-drained mix. A good formula is two parts each of sand and compost to one part of loam. Moisten the mix, sow the seed evenly, then cover lightly with a thin layer of sand, which holds the seed in place. Store in dappled shade in a warm place, and keep damp. Germination usually takes place within 3 weeks. The seedlings can be planted out when big enough to handle, and can be fed with a liquid fertilizer.

Cuttings

Cuttings are very easily rooted, and are best taken during spring or summer. Soft-tip cuttings root very rapidly in moist sand. Semi-hardwood cuttings are generally slower but will easily root. Each cutting should be about 60–100 mm long and have three or four nodes; make the cut just below a node. Keep them in a plastic bag or in water until you are ready to plant them. The medium can consist of sand, perlite, vermiculite, peat or a mixture of any of these. Keep moist and shaded.

Rooting usually takes about two to three weeks in warm weather. The rooted cuttings can be planted out in containers and grown on.

Pests and diseases

Plectranthus are subject to various pests of which the following are commonly encountered. It is better and safer to use biological control agents.

Nematodes

One of the commonest pests is the eelworm or nematode and most *Plectranthus* are affected. These troublesome creatures prevent cell sap from flowing normally, hindering the uptake of minerals and water and the plants suffocate. They are microscopic, and cannot be seen with the naked eye, so plants which suddenly lose vigour must be inspected by uncovering the roots. The characteristic sign is a knobbly thickening. If this is present, take cuttings of the affected plant, and then remove it from the bed, and replace with a nematode-resistant plant. The kind of nematode that attacks *Plectranthus* is only found in the roots so any stem cuttings will be clean. If eelworm is spotted in containerized plants, replace the soil with fresh nematode-free potting soil and re-grow plants from cuttings. One can also treat plants chemically, but most of the effective chemicals are very dangerous and can pollute the environment.

Caterpillars

There are a number of caterpillars that can be troublesome, but the commonest is the *Plectranthus* caterpillar. These are green and so are difficult to spot. They can be removed by hand or sprayed with carbaryl. Plants in pots are generally more vulnerable; in the garden birds keep the numbers down.

Aphids and scale insects (*Aphis* spp.)

These are easily controlled with biological control agents such as ladybirds, gall-midge (*Aphidoletes aphidimyza*) and dwarf wasps (*Aphidius matricaraiae*). Aphids and scale insects can also be controlled by a mineral oil which smothers the scale.

Vine weevil (*Otiorhynchus sulcatus*)

Vine weevils are small hard-bodied insects which leave small round depressions a little larger than a matchhead. These depressions turn brown after feeding. Leaf margins are also often attacked. Biological control agents are available but the plants can also be treated with Carbaspray.

Whitefly (*Trialeurodes vaporariorum*)

Whitefly are seldom as troublesome with plectranthus as they are with pelargonium. The minute wasp *Encarsia formosa* is an effective biological control. The yellow Bugstrip, a sticky card, attracts and traps whiteflies.

Grasshoppers

Grasshoppers sometimes damage the young leaves. In the garden there are usually enough birds and reptiles to keep them at bay. They can be controlled by Carbaspray.

Red spider mite (*Tetranychus cinnabarinus*)

The red spider mite causes discolouring of the surface but is easily controlled by its natural predatory mites (*Phytoseiulus persimilis*, *Amblyseius californicus*). It seems to be troublesome only during hot dry summers and is not a cause of great concern with plectranthus.

Snails

Snails can occasionally be a nuisance. They are active during the night but easily controlled with a shallow jar filled with beer. They are attracted to the irresistible fermenting smell and usually drown themselves in the liquid.

Plectranthus verticillatus is a handy drought-tolerant groundcover. It also does well in a container

4

The indigenous shade garden

Shade adaptations

If one is making a shade garden with plectranthus, it is very helpful to understand how all plants that naturally grow in the shade of forests or bushveld – whether they are trees, shrubs, herbs, ferns or bulbs – manage to survive. Above ground these plants have to cope not only with low light intensities, but with competition from neighbouring plants for light and space. Below ground there is root competition from trees which dries out the soil, and in addition some defences are needed as well against forest-dwelling animals.

The adaptations are many. To conserve moisture, plants may have shallow fleshy roots, fleshy or leathery leaves (which may be divided or incised), fleshy stems or bulbous rootstocks in which they can store water. The root systems may be shallow to take advantage of the upper layer which is rich in leaf litter, and where there is less competition from tree roots (which seek out the deeper layers of permanent moisture below), but which may dry out periodically. The leaves tend to be dark green and smooth rather than grey (which reflects light) or densely hairy, features associated with sunny conditions. The leaf undersides are sometimes suffused with purple, which is caused by the presence of anthocyanins that trap light, ensuring that nothing can travel through the leaf. Many shade plants can adjust the orientation of their foliage, or can increase their leaf size. Most have large leaves to compensate for poor light levels. Some, such as the trailing plectranthus, are mobile and able to move to a spot with better light.

The forest plants fall into three groups. There are the monocots, cycads and ferns which tend to have elongated leaves arranged in rosettes to maximize

The enchanted forest garden at Kirstenbosch where many *Plectranthus* species are grown along with other shade plants such as *Clivia miniata*

1 The autumn-flowering falling stars (*Crocosmia aurea*) makes an ideal companion for blue or white-flowered plectranthus in the shade garden

2 *Veltheimia bracteata* is a drought-tolerant shade-loving bulbous species with striking aloe-like flowers. It is evergreen in warmer areas, and multiplies to form dense clusters that are easily divided

1

light-gathering. This group includes the dracaenas, sansevierias, clivias, and dietes. The second group consists of the creeping, mat-forming herbs whose leaves are characteristically egg-shaped or elliptical. These plants generally have the ability to root at the nodes, and they include many plectranthus and crassulas. The last group includes the taller, branched, sun-loving shrubs and trees. Many of these are fast-growing pioneers that are quick to take advantage of any sudden change in the light caused by a fallen tree, fire or bushpig disturbance. Plants such as *Plectranthus fruticosus*, *P. ecklonii*, the pistol bush (*Duvernoia adhatodoides*) or *Sparrmannia africana* belong here.

Because so many forest plants have large leaves, they often need chemical defences too against forest mammals and other animals. Plectranthus have aromatic leaves and stems, while the forest-dwelling *Zantedeschia albomaculata* as well as its grassland relatives have calcium oxalate crystals in their vegetative parts which make them unpalatable.

Creating a shade garden

All gardens begin sunny, and sooner or later develop shady areas as trees and shrubs increase their canopy. Soon grasses and sun-loving plants are deprived of light and gradually die off. If the leaves and branches are carted away, the result can be bare, dry earth. The surface roots of trees add to the problem by drying out the soil, and the only advantage seems to be that there is less frost in the shade. Fortunately, for every shady place there is be a wealth of indigenous plectranthus and other shade-adapted plants; a change of mindset is all that is needed.

2

The trees. Most people want an instant garden and demand fast-growing shrubs and trees for colour and shade sooner rather than later. The problem is that fast-growing trees have a short lifespan, but if we follow nature, and look to the natural succession of trees in the forest, we can have the best of both worlds. After a tree-fall or forest fire the first trees to germinate are the quick pioneers such as pigeon wood (*Trema orientalis*), keurboom (*Virgilia oroboides*, *V. divaricata*), wild peach (*Kiggelaria africana*), and Cape beech (*Rapanea melanophloeos*). Within a short time their dappled shade provides ideal germinating conditions for seeds of the slower growing species. Under their protective canopy the more permanent yellowwoods (*Podocarpus latifolius*, *P. falcatus*, *P. henkelii*) and Cape chestnut (*Calodendrum capense*) thrive and reach maturity. In creating a forest garden it is therefore best to establish the fast pioneers together with the slower trees. By the time the pioneers are dying the others are large enough to look after themselves. We have used this natural succession to create an artificial forest at Kirstenbosch, and it works! Another excellent pioneer is the sweethorn (*Acacia karoo*) which is friendly to other plants growing in its shade. Other more permanent trees suitable for gardens include the forest mahogany *Trichilia dregeana*, *Schefflera umbellifera*, *Homalium dentatum*, *Apodytes dimidiata*, *Olinia ventosa*, *Nuxia floribunda*, *Vepris reflexa* and *Ficus craterostoma*.

Soil improvement. The first step in making a shade garden is not to remove the leaf litter. The upper layer of the natural forest is rich in leaf debris, which acts as a slow release fertilizer. Organisms in the ground break down the leaves and other plant matter, turning them into nutrient-rich humus which the plants can use. These organisms need moisture to work efficiently, which is why it is so important to keep the moisture-retaining leaf layer, and one can enrich it by adding compost, leaf mould or chipped bark. Even old tree stumps contribute to the health of the garden and should be left to decay. All this organic matter will also help acidify the soil, to provide the slightly acid soils in which most forest plants thrive, though many will tolerate soils of a higher pH. An organic fertilizer will increase the speed of decomposition and make more food available.

Embankments and steep shady slopes. Any steeply sloping ground must be handled with care to avoid soil erosion. It is best planted with shade-tolerant plants which are good soil binders such as hen-and-chickens (*Chlorophytum comosum*) and forest onion (*Drimiopsis maculata*). Embankments should preferably be terraced with dry stone walls or tree trunks to avoid erosion and conserve water.

Establishing plants. The best time to plant is either spring or autumn, preferably at the start of the rainy season. The upper layer of soil should be loosened, and compost and an organic fertilizer such as bonemeal or Neutrog added. Initial watering is crucial, and water should be given until the plants are well established by which time they should be able to tolerate seasonally dry conditions.

Plectranthus companions for dry and moist shade

Dry shade. In the winter-rainfall regions especially the soil in summer becomes very dry. Apart from the plectranthus mentioned in the previous chapter, other plants that can tolerate dry shade are the groundcovering forest crassula, *Crassula multicava*, as well as *Crassula spathulata*, *Clivia miniata*, wild asparagus (*Asparagus densiflorus*), forest sedge (*Cyperus albostriatus*) and *Veltheimia bracteata*. Accent plants such as *Dracaena aletriformis*, wild iris species (*Dietes grandiflora* and others), wild banana (*Strelitzia nicolai*), crane-flower (*Strelitzia reginae*) and the forest cycad (*Encephalartos villosus*) will also cope, as will the seven-weeks fern (*Rumohra adiantiformis*).

For very hot dry bushveld conditions, the succulent mother-in-law's tongues are ideal (*Sansevieria aethiopicum*, *S. hyacinthoides*, *S. concinna* and *S. metallica*). The spotted leaved aloes (*Aloe greenii* and relatives) are good at forming carpets below bushveld trees. The bushveld orchid *Eulophia petersii* also makes groups under trees. The branched arborescent bushveld dragon tree (*Dracaena transvaalensis*) is an ideal accent plant.

Moist shade. Some shade plants require constant moisture and if your garden is in a high-rainfall region, there is a large number of suitable forest species to accompany your plectranthus plantings, such as the many streptocarpus, ferns and impatiens (though some such as *Impatiens zombensis* and *I. hochstetteri*, although beautiful, tend to take over).

Different kinds of shade-adapted plants

Shrubs. The fast-growing pioneer species will make an effect in a very short time. The ribbon bush (*Hypoestes aristata*) looks good in a shady border and can also be massed. Larger species such as the pistol bush (*Duvernoia adhatodioides*) and the Cape stock-rose (*Sparrmannia africana*) will grow tall enough to block out unsightly views. Larger long-lived shrubs include the bird berry (*Psychotria capensis*), Pondo rose-apple

The bird berry (*Psychotria capensis*) makes a beautiful rounded shade-loving shrub to about 3 m. In early summer the yellow flowers attract butterflies, while the red berries that follow are sought after by fruit-eating birds

(*Memecylon grandiflorum*), mackaya (*Mackaya bella*), and wild pomegranate (*Burchellia bubalina*). Good smaller long-lived shrubs are the forest bells (*Mitriostigma axillare*) and forest num-num (*Carissa wyliei*).

Accent plants. Accent plants lighten up dull corners and make the head turn. Their bold shapes or strong features are useful in creating a tropical atmosphere. Suitable plants include the large-leaved strelitzia species (*S. nicolai, S. alba* and *S. caudata*), the large and small-leaved dragon trees (*Dracaena aletriformis* and *D. reflexa* respectively), and the indigenous palms, which also tolerate shade (*Jubaeopsis cafra, Phoenix reclinata*). Many cycads, though not the grey or blue-leaved species, can be used in shade, the most suitable being *Encephalartos villosus, E. natalensis* and *E. paucidentatus*. The smaller fern-cycad (*Stangeria eriopus)* tolerates dry shade. The tree aloe (*Aloe barberae)* will also take shade, but can become very large, though there is a smaller form of it, known as the Tonga tree aloe, which is a better choice for smaller gardens.

Groundcovers. Groundcovers usually have good soil-binding properties, and make a green carpet below trees. The trailing plectranthus species spread by rooting at the nodes, while grasslike plants like the forest sedge (*Cyperus albostriatus*) spread by stolons. The hen-and-chickens (*Chlorophytum comosum*) and bosplakkie (*Crassula multicava*) form plantlets on the inflorescence which root when the flower stem makes contact with the ground. The clivias, especially *Clivia miniata*, are also most useful but slower to multiply. They do bind the soil, however. All these tolerate dry shade, and most are very easily propagated by division and replanted straight back. *Asparagus densiflorus* has fleshy roots which make it very drought-tolerant.

Perennials and bulbs. A number of perennials other than plectranthus tolerate shade. They include *Knowltonia capensis, Gerbera cordata* and *Thunbergia natalensis*. Once established, they are relatively long-lived. There are also many forest-adapted bulbous species, such as the cluster-forming forest crinum (*Crinum moorei*), forest powder brush (*Scadoxus multiflorus* subsp. *katherinae, S. membranaceus, S. puniceus*), falling stars (*Crocosmia aurea*) and *Freesia laxa*.

Climbers. Forest climbers are useful for covering dead shrubs and can be used on shady fences. *Asparagus falcatus, A. plumosus* and *Jasminum multipartitum* are all excellent to plant.

Ferns. Ferns tolerate lower light than flowering plants, and can be used in the darker areas. They vary from the single-stemmed tree ferns to the smaller maidenhair ferns. All generally require year-round moisture. The seven-weeks fern (*Rumohra adiantiformis*) can tolerate dry conditions to some extent.

Epiphytes. Various epiphytic orchids grow well on tree bark in shady conditions. The largest is the Mopane orchid (*Ansellia africana*) which can be established on a tree branch. They are very slow growers but eventually huge and long-lived, and once established, need little care. They feed off the accumulation of debris around their roots, but extra feeding would certainly increase their performance.

The shade-loving weeping bride's bush (*Pavetta lanceolata*) flowers during midsummer

Plants for the forest garden

Shrubs

Name	Habit	Height (m)	Flower colour	Flowering time	Growth rate
Bachmannia woodii	erect	1.5–3	pink	4–8	medium
Brillantaisia subulugurica	erect	2	blue	12–4	fast
Carissa bispinosa	spreading	1–4	white	8–1	medium
Carissa edulis	spreading	1–3	white	9–12	medium
Carissa macrocarpa	spreading	1–4	white	7–11	medium
Carissa wyliei	spreading	1–2	white	10–12	slow
Cassinopsis ilicifolia	spreading	1–4	white	9–11	medium
Cassinopsis tinifolia	spreading	2–7	cream	10–4	medium
Catunaregam spinosa	rounded	3–7	white	8–11	slow
Cephalanthus natalensis	spreading	2–4	greenish	7–11	medium
Choristylis rhamnoides	spreading	1–3	yellow green	8–10	medium
Cryptocarya wyliei	rounded	2–5	white	12–4	medium
Cussonia arenicola	erect	1	yellow-green	11–1	medium
Cussonia natalensis	erect	3–6	yellow-green	4–10	medium
Cussonia nicholsonii	erect	3–5	yellow-green	2–10	medium
Cussonia zuluensis	erect	3–6	yellow-green	10–2	medium
Diospyros natalensis	erect	3–6	white	7–1	slow
Diospyros scabrida	spreading	2–7	cream	10–3	medium
Diospyros whyteana	erect	4–8	cream	8–11	medium
Duvernoia adhatodoides	erect	3–5	white	2–7	fast
Erythroxylon emarginatum	spreading	3–6	green	10–11	medium
Erythroxylon pictum	spreading	1–13	cream	10–2	medium
Hyperacanthus amoenus	spreading	2–7	white	11–3	slow
Indigofera jucunda	spreading	2–4	pink	1–2	fast
Indigofera natalensis	erect	2–4	mauve	2–5	fast
Justicia capensis	spreading	0.2–2	mauve	1–12	medium
Lauridia tetragona	spreading	3–5	cream	10–4	medium
Mackaya bella	erect	1–4	mauve	9–11	medium
Memecylon bachmannii	erect	2–3	greenish	11	medium
Memecylon natalense	erect	8–15	cream	8–11	medium
Micrococca capensis	spreading	1.5–3	greenish	3–5	fast
Mitriostigma axillare	erect	1	white	9–11	medium
Myrsine africana	erect	1	pink	10–12	medium
Ochna natalitia	spreading	1–10	yellow	9–12	slow
Ochna serrulata	spreading	1–3	yellow	9–11	slow
Orthosiphon labiatus	erect	1	purple	2–4	fast
Oxyanthus latifolius	erect	4	white	11–1	medium
Oxyanthus pyriformis	spreading	4–12	white	9–1	medium

Carissa edulis

Indigofera jucunda

Mackaya bella

Ochna serrulata

Shrubs cont.

Name	Habit	Height (m)	Flower colour	Flowering time	Growth rate
Oxyanthus speciosus subsp. *gerrardii*	spreading	4–12	white	11–1	medium
Pavetta gardeniifolia	spreading	2–4	white	11–1	medium
Pavetta gerstneri	spreading	3–5	white	11–12	medium
Pavetta lanceolata	spreading	4–7	white	11–1	medium
Pavetta natalensis	spreading	3–4	white	12–2	medium
Piper capense	shrub	1–3	white	7–4	medium
Psychotria capensis	shrub	3–5	yellow	7–3	medium
Rhamnus prinoides	spreading	1–4	yellow-green	10–11	medium
Rhus dentata	shrub	3–6	yellow	9–11	medium
Sclerochiton apiculatus	spreading	1–3	purplish	11–4	medium
Sclerochiton harveyanus	spreading	2–5	blue	12–7	medium
Sparrmannia africana	spreading	1–3	white	6–11	fast
Tricalysia africana	spreading	2–5	white	8–10	medium
Tricalysia capensis	spreading	2–5	white	11–3	medium
Tricalysia lanceolata	spreading	3–7	white	7–10	medium
Tricalysia sonderiana	spreading	2–5	white	9–12	medium
Trichocladus crinitus	spreading	2–4	yellow	8–10	medium
Trichocladus ellipticus	spreading	3–7	yellow	9–12	medium
Trichocladus grandiflorus	spreading	7–15	white	10–12	medium

Psychotria capensis

Sparrmannia africana

Accent and container plants

Name	Habit	Height (m)	Flower colour	Flowering time	Growth rate
Aloe barberae	erect	5–15	pink	6	fast
Aloe pluridens	erect	2–5	orange or yellow	5–6	medium
Cussonia arenicola	erect	1	yellow-green	11–1	medium
Cussonia nicholsonii	erect	3–5	yellow-green	4	medium
Cussonia zuluensis	erect	3–4	yellow-green	11–1	medium
Dracaena aletriformis	erect	2–4	white	12–2	medium
Dracaena mannii	spreading	3–5	cream	3–5	medium
Encephalartos altensteinii	erect	2–5	yellow-green	11–3	slow
Encephalartos natalensis	erect	3–5	yellow-green	11–3	slow
Encephalartos paucidentatus	erect	3–6	yellow-green	11–3	slow
Encephalartos transvenosus	erect	4–10	yellow-green	11–3	slow
Encephalartos villosus	erect	1–2	yellow-green	11–3	slow
Ensete ventricosum	erect	4–5	red-brown	10–11	fast
Jubaeopsis caffra	erect	3–8	white	9–10	slow
Sclerochiton harveyanus	spreading	2–5	blue	12–7	medium
Stangeria eriopus	rosette	1–1.5	yellow-green	5–10	slow

Encephalartos paucidentatus

Name	Habit	Height (m)	Flower colour	Flowering time	Growth rate
Strelitzia alba	erect	4–10	white	10–2	medium
Strelitzia caudata	erect	4–10	white	1–3	medium
Strelitzia nicolai	erect	4–10	white	10–2	medium
Strelitzia reginae	spreading	1–1.5	orange	10–2	slow

Herbaceous perennials

Name	Shape	Height (m)	Flower colour	Flowering time	Growth rate
Asystasia gangetica	spreading	0.2	white	1–5	fast
Asystasia varia	spreading	0.2	mauve	3–5	medium
Barleria repens	spreading	0.3	red	2–5	medium
Begonia dregei	spreading	0.3	pink	12–8	medium
Begonia homonyma	spreading	0.3	pink	12–3	medium
Begonia sonderana	spreading	0.5	pink	12–4	fast
Begonia sutherlandii	spreading	0.3	orange	11–4	fast
Coleotrype natalensis	tufted	0.5	purple	9–4	medium
Gerbera cordata	tufted	0.2	white	10–4	medium
Hypoestes aristata	erect	1.5	mauve	3–8	medium
Impatiens flanaganiae	erect	0.6	pink	11–12	fast
Impatiens hochstetteri	spreading	0.3	pink	11–5	fast
Impatiens sylvicola	spreading	0.2	pink	11–5	fast
Impatiens walleriana	spreading	0.2	pink, mauve	11–3	fast
Impatiens zombensis	spreading	0.3	pink, mauve	11–3	fast
Isoglossa cooperi	spreading	1.5	pink	3–5	fast
Isoglossa woodii	spreading	1–4	white	3–7	fast
Justicia betonica	shrublet	0.3–1	white	4–12	fast
Justicia campylostemon	shrublet	1–3	white	12–5	fast
Justicia petiolaris subsp. *bowiei*	shrublet	1	mauve	12–5	fast
Justicia protracta	shrublet	0.3–1	white	9–7	fast
Knowltonia capensis	tufted	0.2	green	2–4	medium
Laportea grossa	herb	0.5	white	2	fast
Lobelia cuneifolia	spreading	0.3	blue	11–5	fast
Pelargonium tongaense	spreading	0.2	red	11–4	fast
Streptocarpus caeruleus	tufted	0.1	violet, blue	12–4	fast
Streptocarpus cyaneus	tufted	0.1	violet, white	10–5	fast
Streptocarpus gardenii	tufted	0.1	light blue	12–4	fast
Streptocarpus meyeri	tufted	0.1	pale violet	11–1	fast
Streptocarpus rexii	tufted	0.1	blue	12–4	fast
Thunbergia natalensis	shrublet	1.2	mauve	10–3	fast

Asystasia gangetica

Hypoestes aristata

Thunbergia natalensis

Streptocarpus cultivar

Crassula multicava

Groundcovers

Name	Leaf colour	Flower colour	Flowering time	Growth rate	Light
Asparagus densiflorus 'Cwebe'	green	white	2–5	medium	light shade
Asparagus densiflorus 'Mazeppa'	green	white	2–5	medium	light shade
Asparagus oxyacanthus	green	white	11–5	medium	light shade
Asparagus ramosissimum	green	white	9–2	medium	light shade
Chaetacanthus setiger	green	white	3–5	medium	light shade
Chlorophytum comosum	green	purple	9–10	fast	shade
Crassula inandensis	shiny	cream	11–1	slow	shade
Crassula multicava	green	white	1–5	medium	shade
Crassula pellucida subsp. *marginalis*	green	white	9–12	fast	shade
Crassula spathulata	green	white	3–5	fast	shade
Crassula streyi	green	white	5–7	slow	shade
Cyperus albostriatus	green	yellow-green	1–4	medium	shade
Dichondra repens	green	greenish	9–5	fast	shade
Dicliptera clinopodia	green	purple	3–5	fast	shade
Diclis reptans	green	white	6–7	fast	shade
Drimiopsis maculata	mottled	white	11	medium	shade
Ledebouria socialis	mottled	purple	10–3	medium	shade
Monopsis lutea	green	yellow	9–5	fast	shade
Monopsis unidentata	green	purple	9–4	fast	light shade
Oplismenus hirtellus	green	green	1–7	medium	shade
Rhoicissus rhomboidea	green	green	11–2	medium	light shade
Senecio bryoniifolius	green	yellow	5	fast	shade
Stachys tubulosa	green	mauve	0.4	fast	shade

Clivia gardenii

Bulbous plants

Name	Habit	Size (m)	Flower colour	Flowering time	Growth rate
Agapanthus praecox subsp. *orientalis*	tufted	1	blue/ white	12–2	fast
Bonatea lamprophylla	tufted	1.2	white	9–10	medium
Bonatea porrecta	tufted	0.6	white	7–8	medium
Bonatea speciosa	tufted	1	white	7–2	medium
Brownleea coerulea	tufted	0.6	mauve	2–4	medium
Calanthe sylvatica	tufted	0.7	mauve	11–3	medium
Chlorophytum bowkeri	tufted	1	white	12–4	medium
Chlorophytum krookianum	tufted	1–2	white	12–4	medium
Clivia caulescens	tufted	0.6	reddish	9–3	slow
Clivia gardenii	tufted	0.6	reddish	9–12	slow

Name	Habit	Size (m)	Flower colour	Flowering time	Growth rate
Clivia miniata	tufted	0.6	red/ orange/ yellow	8–10	slow
Clivia nobilis	tufted	0.6	reddish	8–10	slow
Crinum moorei	tufted	1	pink	11–3	medium
Crocosmia aurea	erect	0.6	orange	3–4	fast
Crocosmia paniculata	erect	1.8	orange	1–2	fast
Dietes bicolor	tufted	0.8	white	10–3	medium
Dietes butcheriana	tufted	0.6	white	11–2	slow
Dietes flavida	tufted	0.7	yellow	9–4	medium
Dietes grandiflora	tufted	1	white	11–1	medium
Dietes iridioides	tufted	0.6	white	8–4	medium
Dietes vegeta	tufted	0.7	yellow	10–3	medium
Drimiopsis maculata	tufted	0.01	white	10–11	medium
Eulophia horsfallii	tufted	2	pink	9–7	medium
Gloriosa superba	climbing	1–2	yellow/ orange	12–5	medium
Gonatopus boivinii	tufted	1.2	mottled	11–12	medium
Haemanthus albiflos	tufted	0.2	white	2–4	slow
Haemanthus deformis	tufted	0.2	white	11–2	slow
Haemanthus humilis	tufted	0.1	white	1–2	slow
Scadoxus membranaceus	tufted	0.4	orange	1–5	medium
Scadoxus multiflorus subsp. *katharinae*	tufted	0.6	orange	11–1	medium
Schizostylis coccinea	tufted	0.7	red	11–4	medium
Siphonochilus aethiopicus	tufted	0.5–1	pink	10–2	medium
Stenoglottis fimbriata	tufted	0.4	pink	1–4	medium
Stenoglottis longifolia	tufted	0.4	pink	5–7	medium
Stenoglottis woodii	tufted	0.2	pink	10–5	medium
Veltheimia bracteata	tufted	0.2	orange	7–9	medium
Zamioculcas zamiifolia	tufted	0.8	yellowish	12–3	medium
Zantedeschia aethiopicum	tufted	0.6	white	8–2	medium
Zantedeschia albomaculata	tufted	0.7	white	11–3	medium
Zantedeschia rehmannii	tufted	0.6	pink/ maroon	11–12	medium

Clivia miniata* var. *citrina

Scadoxus membranaceus

Scadoxus multiflorus* subsp. *katharinae

***Veltheimia bracteata* 'Lemon Flame'**

Climbers

Name	Habit	Height (m)	Flower colour	Flowering time	Growth rate
Acridocarpus natalitius	twiner	1–15	yellow	11–2	slow
Adenia fruticosa	climber	2–8	white	9–11	medium
Adenia gummifera	climber	5–15	white	10–12	fast

Acridocarpus natalitius

Climbers cont.

Name	Habit	Height (m)	Flower colour	Flowering time	Growth rate
Asparagus falcatus	climber	2–5	white	10–2	medium
Asparagus macowanii subsp. *zuluensis*	climber	1–2.5	white	10–11	fast
Asparagus plumosus	scandent	2–3	white	2–3	fast
Behnia reticulata	scandent	2–5	white	1–12	medium
Cussonia thyrsiflora	scrambler	1–5	greenish	11–1	medium
Cyphostemma rubroglandulosum	climber	3–10	green	11–2	fast
Cyphostemma simulans	climber	3–8	green	11–2	medium
Dioscorea cotinifolia	twiner	1–10	greenish	10–1	slow
Dioscorea dregeana	twiner	1–15	white/ green	10–1	slow
Dioscorea retusa	twiner	1–4	greenish	12–1	slow
Dioscorea rupicola	twiner	1–3	greenish	11–1	slow
Dioscorea sylvatica	twiner	1–4	greenish	11–4	slow
Diospyros villosa	scrambler	1–4	cream	3–5	medium
Flagellaria guineensis	climber	3–10	cream	10–1	fast
Gerrardanthus macrorhizus	climber	1–5	orange	10–4	medium
Gloriosa superba	climber	1–1.5	orange	11–3	fast
Jasminum abyssinicum	scandent	8	white	3–5	medium
Jasminum multipartitum	climber	1–3	white	9–1	medium
Jasminum stenolobum	climber	1–5	white	9–1	medium
Jasminum streptopus var. *transvaalensis*	scandent	1–3	white	10–1	medium
Petopentia natalensis	twiner	1–10	maroon	9–11	slow
Senecio bryoniifolius	climber	1–5	yellow	5	medium
Senecio macroglossus	twiner	1–3	yellow	9–5	medium
Senecio quinquelobus	twiner	1–3	yellow	4–7	medium
Senecio tamoides	twiner	1–10	yellow	3–5	fast
Thunbergia alata	twiner	2–3	orange	9–5	fast

Senecio tamoides

Thunbergia alata

Tufted perennials (grasses, sedges, restios)

Name	Shape	Size (m)	Flower colour	Flowering time	Growth rate
Cyperus albostriatus	tufted	0.4	green	10–7	fast
Olyra latifolia	tufted	0.6	green	9–2	slow
Schoenoxiphium lanceum	tufted	0.6	green	6–11	medium
Setaria megaphylla	tufted	0.15–0.2	green	9–7	fast

Epiphytes

Name	Habit	Flower colour	Flowering time	Growth rate
Acampe praemorsa	groups	yellow	11–3	slow
Aerangis mystacidii	groups (dwarf)	white	2–3	slow
Ansellia africana	clustered	yellow	7-8	slow
Bulbophyllum longiflorum (Cirrhopetalum umbellatum)	mat-forming	flesh	10–1	slow
Bulbophyllum sandersonii	mat-forming	maroon	10–12	slow
Bulbophyllum scaberulum	mat-forming	pale yellow	10–12	slow
Cyrtanthus epiphyticus	tufted	red	11–2	slow
Cyrtorchis arcuata	decumbent groups	white	1–2	slow
Cyrtorchis praetermissa	groups	white	11–12	slow
Dermatobotrys saundersiae	herbaceous to 30 cm	orange	7–12	medium
Diaphananthe xanthopollinia	pendulous	white	2	medium
Microsorum scolopendria	creeping	spores	1–12	medium
Peperomia blanda	groups	greenish	1–7	medium
Peperomia retusa	groups	greenish	12–4	fast
Peperomia rotundifolia	mat-forming	greenish	1–9	medium
Peperomia tetraphylla	groups	greenish	1–9	medium
Pleopeltis macrocarpa	creeping	spores	1–12	slow
Polystachya ottoniana	mat-forming	white/ pale green	7–12	slow
Polystachya pubescens	tufted	yellow	10–12	slow
Polystachya zambesiaca	clustered	white	10–12	slow
Tridactyle bicaudata	clustered	white	10–2	slow
Vanilla roscheri	scandent	white	1	medium
Zamioculcas zamiifolia	herbaceous, to 0.8 m	yellowish	12–3	medium

Ansellia africana

Peperomia retusa

Ferns

Name	Habit	Height (m)	Time	Rate
Adiantum capillus-veneris	creeping	0.3	8–5	medium
Adiantum poiretii	creeping	0.5	7–4	medium
Asplenium aethiopicum	creeping	0.5	7–8	medium
Asplenium lunulatum	tufted	0.4	10–12	medium
Asplenium rutifolium	epiphytic	0.3	12–8	medium
Blechnum attenuatum	tufted	1	11–7	slow
Cyathea capensis	tree fern	1–3	10–7	slow
Cyathea dregei	tree fern	1–4	6–7	slow
Cyrtomium caryotideum	erect	0.7	1–12	medium

Cyathea dregei

Ferns cont.

Stenochlaena tenuifolia

Name	Habit	Height (m)	Time	Rate
Davallia chaerophylloides	groundcover	30	4–5	medium
Didymochlaena truncatula	tufted	1	1–12	slow
Diplazium zanzibaricum	tufted	1	7	medium
Marattia fraxinea	tufted	2	1–12	slow
Microsorum scolopendria	creeping	60 cm	1–12	slow
Osmunda regalis	tufted	1	1–12	slow
Rumohra adiantiformis	creeping	0.5	1–12	slow
Selaginella kraussiana	procumbent	0.05	1–12	fast
Stenochlaena tenuifolia	climber	3–5	11–3	medium
Tectaria gemmifera	tufted	1	10–2	medium
Thelypteris bergiana	tufted	0.5	8–5	fast
Todea barbara	tufted	0.1	1–12	slow

Succulent plants for dry shady bushveld gardens

Aloe barberae

Aloe pluridens

Name	Habit	Height (m)	Flower colour	Flowering time	Growth rate
Aloe barberae	treelike	5–10	pink/ orange	8–9	medium
Aloe greenii	groups	0.3	orange	1–3	medium
Aloe mudenensis	groups	0.3	red/ orange	7	medium
Aloe parvibracteata	groups	0.3	red	6–7	medium
Aloe pluridens	erect shrub	3	orange	5	slow
Aloe rupestris	erect shrub	3	orange	6–7	slow
Aloe suffulta	decumbent	0.3	red	7–3	slow
Aloe umfoloziensis	groups	0.7	orange	7–8	medium
Aloe vanbalenii	groups	0.6	red/ yellow	6–7	medium
Cyphostemma cirrhosum	climbing	1–10	yellowish	10–12	slow
Cyphostemma hypoleucum	climbing	1–2	yellowish	10–5	slow
Gasteria acinacifolia	tufted	0.6	pink	8–11	slow
Gasteria batesiana	tufted	3	orange	11–2	slow
Gasteria croucheri	tufted	3	orange	11–2	slow
Gasteria excelsa	solitary	3	orange	11–1	slow
Kalanchoe crenata	shrublet	0.2–1	yellow-orange	5–8	medium
Sansevieria concinna	tufted	0.5	white	1–2	slow
Sansevieria hyacinthoides	tufted	0.6	white	1–2	slow

5

Plectranthus cultivars and their breeding potential

Most garden plants originate through a process of selection. Whether one is selecting a particular plant from a wild population, or whether one selects a particular seedling arising from a deliberate breeding programme, the principle is the same. Once the selection is made, the plant may be given a cultivar name. The word 'cultivar' means a 'cultivated variety', and the only requirement is that the plant has to be in cultivation. A cultivar name can be given to mutations, hybrids, selections from the wild, or colour variants, but never to a plant still growing in the field. The naming of wild plants is still governed by the strict rules of the International Code of Botanical Nomenclature.

There are various ways of producing plectranthus cultivars. One can select good forms with superior qualities from the wild, or even from the garden; another

In this nursery *Plectranthus hadiensis* var. *tomentosus* dominates the foreground, with the *P. ecklonii* cultivars 'Erma' behind to the right and 'Tommy' at the back

is to engage in a focused breeding programme with a definite aim in mind. Yet another (of which I do not approve) is to irradiate plant material to induce genetic mutations such as leaf variegation.

Selection from the field and cultivar naming

I first fell in love with plectranthus when I was a young student working in the Lowveld Botanical Gardens, and I distinctly remember that it was the conspicuously beautiful flowering panicles and ornamental leaves of *P. fruticosus* and *P. ecklonii* which I saw in the cool escarpment forests that turned my head. I soon found, however, that although they rooted rapidly, many did not thrive well in the warmer dry bushveld climate, and I began to realize how important it was to select plants suited to one's climate.

Dr Codd's revision of the group had just been published, and with this as my guide I determined to collect and introduce all the species I could. I soon noticed that there seemed to be local forms and variants in the different sites I visited. The variation differed according to the species. Most species showed a good deal of variation (*P. verticillatus* and *P. fruticosus* were examples) whilst others showed less. Some forms from Barberton growing on the Kimberley Reef geological formation had dark purple undersides whilst others were very hairy. I could thus start selecting what I considered to be the better forms, and that is exactly what I did. When I was transferred to Kirstenbosch I found the cool afrotemperate climate there much better suited to the group, and I started a process of selection in earnest, keeping only the very best for cultivation.

When a plant is introduced from the field it is immediately labelled and all the collection details are recorded. With our plectranthus we decided to give our specially selected plants some easily recognizable cultivar name instead of a number. We named the plants after their locations (if we knew them) or after the people who had contributed in some way. Thus *P. hilliardiae* 'Magwa' was collected at the Magwa Falls in the Eastern Cape where this superior form with its shiny leaves and dark purple undersides was found. A very beautiful long-tubed white flowering form of *P. saccatus* subsp. *saccatus* from Ngoye Forest in KwaZulu-Natal was given the cultivar name 'King Goodwill' to honour the king's contribution to conserving this forest. *P. fruticosus* 'James' was named after the conservationist ecologist James Culverwell who discovered this beautiful pink form on the South African side of the border with Swaziland.

***Plectranthus* 'Mona Lavender', bred by Roger Jacques, is here seen flowering in Gert Brits's Stellenbosch garden**

Selection from the garden

Plectranthus grown in gardens also shows variation, especially when grown from seed. If a plant is notably different – in flower colour, or disease-resistance – it can be given a cultivar name. Occasionally a single branch might display a different feature, such as leaf variegation. This is usually due to natural mutation and if cuttings are made from the branch, they should all display the same character. If they don't, it shows that the variegation was due to soil deficiencies, or virus infections or some other cause. All cuttings made from *Plectranthus madagascariensis* 'Lynne', for instance, will display the beautiful variegation (white, pink and shades of green) of the original mutation. This cultivar is named after Lynne Nichols who painted many of the species when we started our *Plectranthus* project.

Different species grown together in the garden often produce hybrids, and there may well be plants worth naming among these.

Focused breeding and 'Mona Lavender'

In my thirty years of finding and selecting plants, I have always been aware of the breeding potential of *Plectranthus*, and had always hoped that someone would develop the group in the same way that *Pelargonium* was developed. In particular I felt crosses between the large-flowering *Plectranthus saccatus* and *P. hilliardiae* would produce beautiful hybrids. We had many good forms of *P. saccatus* but only a solitary one of *P. hilliardiae* from Mtamvuna, and I needed to find a better form. Eventually I came across some very fine plants at Magwa in the Transkei, growing close to the waterfall. The leaves were elliptical, often purple below and of a beautiful dark shiny green. Within the population there was also variation in leaf shape, the amount of purple below, hairiness, and shiny character. We named the best form 'Magwa' and grew it successfully in our greenhouses at Kirstenbosch.

We then crossed the two species, transferring pollen from *P. saccatus* to 'Magwa', and some seed was set. The plants that germinated that spring were indeed hybrids; they had the beautiful dark purple shiny leaves of *P. hilliardiae*, and showed signs of wanting to flower freely. However they still needed further work. Our newly appointed horticulturist Roger Jacques took up the project, and started a breeding programme with these two species, using 'Magwa' and several forms of *P. saccatus*. He went on many expeditions to the field, and his work eventually led to the release of 'Mona Lavender' in 2002.

Other breeders

Another breeder is Gert Brits, a geneticist by training who has been breeding plants for more than four decades. He has released or helped develop about 50 cultivars of protea, many of which are his own hybrids, and he has also written extensively about their cultivation. He first became interested in *Plectranthus* during his work at Kirstenbosch in the nineties. After seeing *P. hilliardiae* in full flower at Clearwater Mtamvuna, he was seized with the idea of using it to breed a race of house plants. After his retirement from the Fynbos Research Station at Stellenbosch, he started a breeding programme using the Kirstenbosch forms, and his passion can be seen in the beautiful cultivars he made. Using *P. hilliardiae* and *P. saccatus* he created the famous 'Cape Angels' series which was released worldwide in 2001, and which includes the very fine 'Pink Angel' and the lovely 'Blue Angel', originally named 'Edelblau'. 'Cloud Nine' is another very fine and very late-flowering hybrid also derived from *P. hilliardiae.* He also created 'Coral Cloud' from *P. fruticosus;* 'Lilac Spur' and 'Seasheen' from *P. oertendahlii;* the white 'Guru's Choice' from *P. saccatus;* and 'Amanda' from *P. purpuratus.*

Gert Brits inspects a *Plectranthus* cultivar in his garden

Plectranthus breeding is growing steadily here and abroad. Competition is strong as breeders try to achieve plants that will sell well but which are also sterile, so that no one else can use their creations to develop new cultivars. Breeding new cultivars is a time-consuming and expensive process, and breeders do not like to have their breeding lines exposed. Most of the energy has gone into producing house plants, but there is still a large market for developing garden plants. There are enough species and selected cultivars to ensure a great future for any breeder who undertakes to do this.

This cross between *Plectranthus hilliardiae* 'Magwa' and *P. saccatus* 'Nkandla' has not yet been released

Cultivars bred from the Southern African Plectranthus

Cultivar	Parents	Habit	Flowering time	Colour	Breeder
'Athens Gem'	*P. amboinicus*	spreading low-growing groundcover	autumn	white	Allan Armitage's New Crop Breeding Program
'Amanda'	× *P. purpuratus*	spreading low-growing ground cover	summer & autumn	light mauve	Gert Brits
'Cloud Nine'	× *P. hilliardiae*	spreading shrublet	late autumn & winter	blue	Gert Brits
'Cape Angel Blue' (='Edelblau')	*P. hilliardiae* × *saccatus*	perennial shrublet	summer, peaks in autumn	blue	Gert Brits
'Cape Angel Dark Pink'	*P. hilliardiae* × *saccatus*	shrublet	summer, peaks in autumn	deep pink	Gert Brits
'Cape Angel Dark Purple'	*P. hilliardiae* × *saccatus*	shrublet	summer, peaks in autumn	deep purple	Gert Brits
'Cape Angel Pink'	*P. hilliardiae* × *saccatus*	shrublet	summer, peaks in autumn	pink	Gert Brits
'Cape Angel Purple'	*P. hilliardiae* × *saccatus*	shrublet	summer, peaks in autumn	purple	Gert Brits
'Cape Angel White'	*P. hilliardiae* × *saccatus*	shrublet	summer, peaks in autumn	white	Gert Brits
'Coral Cloud'	× *P. fruticosus*	shrublet	autumn	coral pink	Gert Brits
'Easy Gold'	*P. ciliatus*	groundcover, yellowish leaves	autumn	white	unknown
'Frills'	× *P. fruticosus*	shrublet, frilly leaves	autumn	light pink	unknown
'Gold Coin'	*P. ciliatus*	groundcover, yellowish leaves	autumn	white	unknown
'Guru's Choice'	*P. saccatus (white)*	spreading shrublet	autumn	white	Gert Brits
'Lilac Spur'	× *P. oertendahlii*	spreading perennial	autumn	lilac pink	Gert Brits
'Mona Lavender'	*P. hilliardiae* × *saccatus*	shrublet	summer to winter, peaks in autumn	mauve	Roger Jaques
'Nico'	*P. ambiguus*	spreading groundcover	autumn	mauve	Ball Floraplant
'Seasheen'	*P. oertendahlii*	spreading perennial	autumn	white	Gert Brits
'Silver Star'	*P. oertendahlii*	spreading perennial	autumn	white	Van der Voort

1

2

3

All these plants were bred by Gert Brits:

1 'Cape Angel Dark Pink'
2 'Cape Angel White'
3 'Cape Angel Pink'
4 Various new cultivars
5 A new triploid cultivar not yet released

4

5

6 Distribution and ecology

Distribution

Most of the 350 or so *Plectranthus* species of the world are soft trailing semi-succulent to succulent herbs or shrubs widely distributed in the summer-rainfall savannahs and forested regions of Africa, Madagascar, India, Australia and a few Pacific islands. *Plectranthus* appears to be an old world genus absent from South America, though the majority are clearly African. Because the boundaries of the genus are still unclear, we cannot be sure exactly how many species there are. At present various closely related groups (such as *Coleus*, *Burnatastrum* and *Xerophilus*) are regarded as sections within the greater genus *Plectranthus*, but they may well be raised to separate genera. Once the problems have been sorted out it is estimated that *Plectranthus* will encompass about 200 species. Of these about a quarter occur in South Africa where there are 53 known species.

The South African species are clearly confined to the south-eastern and eastern parts, with the majority occurring east of the Drakensberg escarpment. They are found in all the provinces except the Northern Cape,

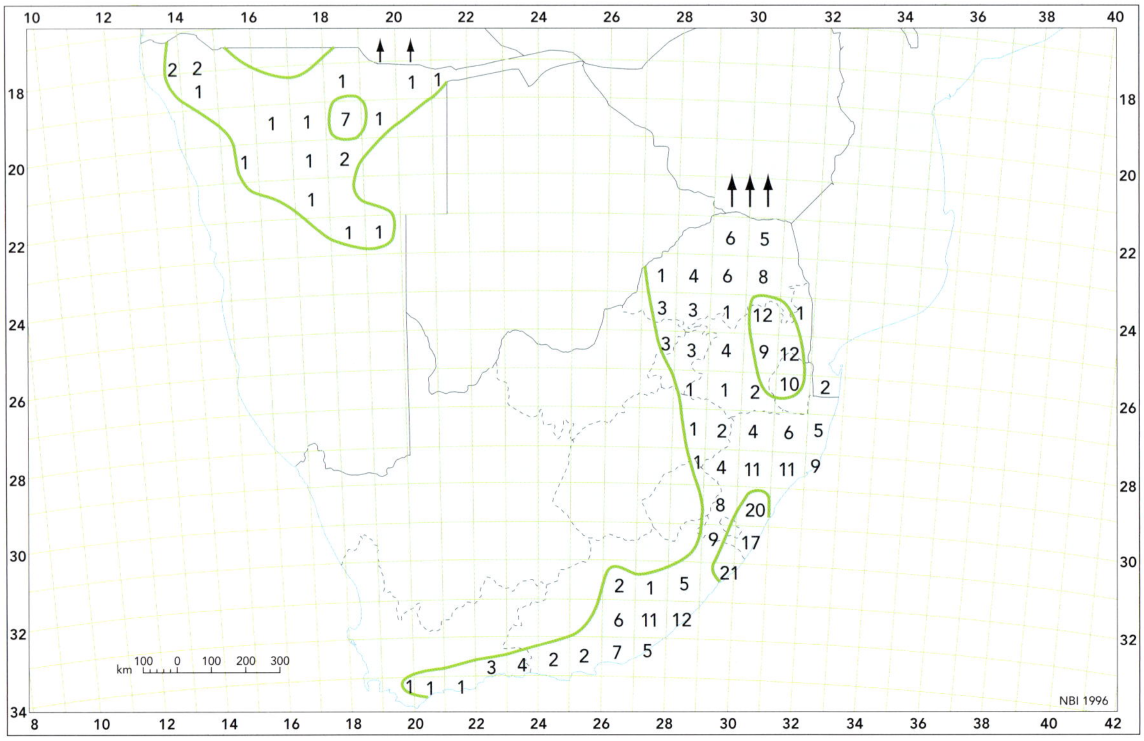

The distribution of *Plectranthus* in Southern Africa showing the number of species in each square degree. The ringed portions refer to centres of *Plectranthus* diversity

and they also occur in northern Namibia, Swaziland and Lesotho. They are rare in the Western Cape and Free State (2 species each). There are two clear centres of diversity: one is in the northern coastal part of the Eastern Cape and the adjacent southern part of KwaZulu-Natal (21 species), and the other is in the Mpumalanga Drakensberg and adjoining Swaziland region (18 species). The reasons for the great diversity here probably have to do with two uplifts of the African continent; an initial minor uplift (to 300 m) in the early Miocene about 20 million years ago, and a more recent major uplift in the south-east Cape (to 900 m) during the Pliocene 3–5 million years ago, which left the landscape full of different and often isolated habitats that encouraged speciation.

Southern Africa has seven major vegetation regions or biomes. *Plectranthus* is rare or absent in the winter-rainfall regions of fynbos and succulent Karoo, and entirely absent from the summer-rainfall inland Karoo. A few species are confined to the higher altitude grasslands of the eastern and northern parts of the country. The numbers increase dramatically, however, in the savannahs, the forests, and the succulent thickets of the eastern river valleys.

1

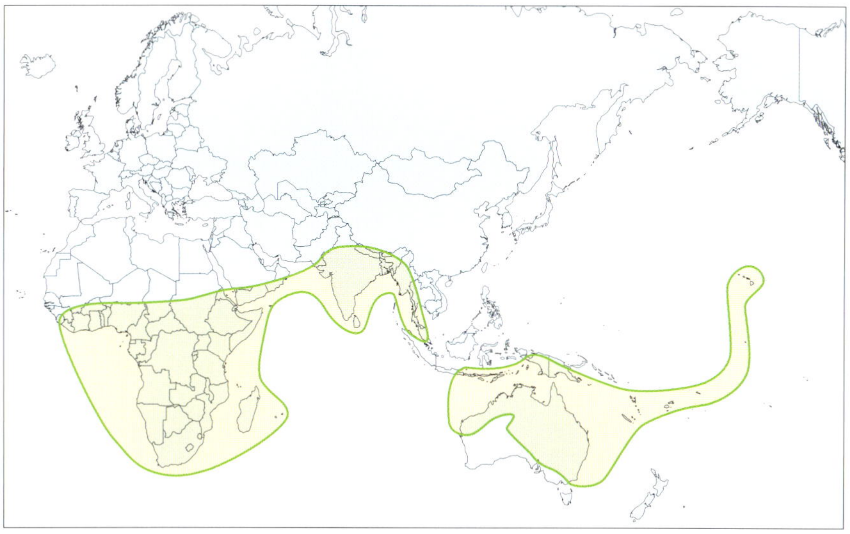

2

1 The world distribution of *Plectranthus*

2 *Plectranthus grandidentatus* growing among *Aloe arborescens* on a rocky scree at the fringe of a cool, moist afromontane forest near Hlobane, KwaZulu-Natal

Forests of the east and south

The Southern African forests make up the smallest surface area of our vegetation regions but are clearly the richest in *Plectranthus* species. There are two types of forest: the southern warm-temperate and sometimes high altitude afromontane forest, and the lowland subtropical forest along the eastern coast. *Plectranthus* are commonly found as ground-layer herbs in both types, or as shrubs on their margins. These forest are evergreen and the annual rainfall usually exceeds 1000 mm. The soil varies according to the rock formation, but is usually weathered sandstone rich in humus which remains damp throughout the year. The forests are also home to various animals such as bush-pig, bushbuck and porcupine, and plants are well adapted to the disturbances they cause. Fires are often frequent along the margins where some of the shrubby species occur.

Afromontane forests

The afromontane forests follow the eastern side of the Drakensberg as far north as the Zoutpansberg, and go south to Knysna and the Tsitsikamma. They are highly dissected, and are usually found on the cooler southern and eastern slopes above 1000 m, where the rainfall is higher. The climate is cool, and the winters are cold but seldom frosty. To the north they are found at progressively higher altitudes along the east African mountain chain.

Eleven species of *Plectranthus* are found in our afromontane forests: the shrubby *P. ecklonii*, *P. fruticosus*, *P. grallatus*, *P. rubropunctatus* and *P. rehmannii*; the smaller species and groundcovers *P. ciliatus*, *P. elegantulus*, *P. grandidentatus*, *P. swynnertonii, P. verticillatus*; and the small shrublet *P. dolichopodus*. Most of these also occur in the subtropical coastal forests and elsewhere, especially *P. fruticosus*, *P. grallatus*, *P. ecklonii*, *P. verticillatus* and *P. ciliatus*. Two occur only in the KwaZulu-Natal midlands: *P. elegantulus* and *P. rehmannii*, which is confined to Karkloof. *P. swynnertonii* is found as well in Limpopo Province and again in the Chimanimani forests in Zimbabwe, while *P. rubropunctatus* is widespread in the forests of the eastern escarpment of Mpumalanga and Limpopo.

Common afromontane trees associated with *Plectranthus* include the broad-leafed yellowwood (*Podocarpus latifolius*), yellowwood (*P. falcatus*), terblans (*Faurea macnaughtonii*), red alder (*Cunonia capensis*), white alder (*Platylophus trifoliatus*), assegai (*Curtisia dentata*), black stinkwood (*Ocotea bullata*), wild peach (*Kiggelaria*

1

2

africana), hard pear (*Olinia ventosa*), Cape beech (*Rapanea melanophloeos*), Cape chestnut (*Calodendrum capense*), ironwood (*Olea capensis*), black ironwood (*O. macrocarpa*), holly (*Ilex mitis*), Cape saffron (*Cassine peragua*), white pear (*Apodytes dimidiata*) and small knobwood (*Xanthoxylon capensis*). Understory plants include witch-hazel (*Trichocladus crinitus*), stock-rose (*Sparrmannia africana*) and wild pomegranate (*Burchellia bubalina*).

Subtropical lowland forests

The lowland forests are rich not only in *Plectranthus* but also in local endemics, with a clear centre of diversity along the sandstone Cape Supergroup chain that runs from southern KwaZulu-Natal through the northern coastal Eastern Cape (the former Transkei). The climate, which is influenced by the warm Indian ocean, is warm, humid and frost-free. Although the forests follow a broad belt along the coast they are patchy and confined to the coastal river valleys. Some river systems are so isolated that species or forms may be confined to them (the Mzimvubu has four endemics, for example). Seventeen *Plectranthus* are found only in these forests. Most widespread are *P. saccatus*, *P. zuluensis*, *P. ecklonii* and *P. fruticosus*, and smaller forest trailers *Plectranthus aliciae*, *P. laxiflorus*, *P. madagascariensis*, *P. petiolaris*, *P. strigosus*, *P. ambiguus*, *P. ciliatus*, *P. purpuratus* and *P. lucidus*, the latter confined to coastal dunes between Port Elizabeth and Port St Johns. The rest are all local endemics. These include the shrubs *P. oribiensis* (confined to the Oribi Gorge), *P. reflexus* (the Mzimvubu) and *P. mzimvubuensis* (the inland Mzimvubu, on shale). Smaller local endemics confined to isolated forested sandstones include *P. malvinus, P. hilliardiae*, *P. praetermissus*, *P. ernstii*, *P. stylesii* and *P. oertendahlii*.

Common trees associated with *Plectranthus* in these lowland forests include the wild plum (*Harpephyllum caffrum*), myrtle quince (*Cryptocarya myrtifolia*), broad-leaved quince (*Cryptocarya latifolia)*, Natal quine (*Dahlgrenodendron natalense*), forest mahogany (*Trichilia dregeana*), *Protorhus longifolius*, the broad-leafed yellowwood (*Podocarpus latifolius*), yellowwood (*P. falcatus*), dwaba-berry (*Monanthotaxis caffra*), Bizana fig (*Ficus bizanae*), polished fig (*Ficus polita*), forest fig (*F. craterostoma*), onion-wood (*Cassipourea gummiflua*), forest bush-willow (*Combretum kraussii*) large-leaved myrtle (*Eugenia erythrophylla*), Cape holly (*Ilex mitis*) and Natal coast cabbage tree (*Cussonia nicholsonii*). Understory plants include *Dracaena hookeriana*, *Flagelaria guineensis*, *Streptocarpus* spp., *Chlorophytum comosum*, *Crassula streyi*, and *C. inandensis*.

Grassveld of the eastern highlands

There are very few grassland *Plectranthus* in the vast regions of grassveld in the east. The terrain is very variable (flat to mountainous) and the geology diverse (rocks are sedimentary to igneous). Soils can be sandy or clayey. Grasslands are subject to regular disturbances, which include fire, grazing animals and winter frost; the rainfall also varies greatly. Grassland-adapted species have perennial tuberous rootstocks which resprout annually during spring. *Plectranthus ramosior* occurs in rocky grassland on quartzitic soil in the Gauteng, North-West and Limpopo Provinces, and the related *P. woodii* and *P. hadiensis* var. *tomentosus* in the east from the Eastern Cape to Mpumalanga.

Bushveld (savannah) and succulent thicket

The bushveld, which belongs to the savannah biome, occurs in the north and covers a very large part of South Africa. After the forests, it has the next highest occurrence of *Plectranthus* (21 species). The vegetation is grassland with scattered trees and shrubs on flat or hilly and rocky terrain. The geology can be very complex and soils vary from sand to clay. The climate is subtropical with warm humid summers but the winters are dry and there can be light frost at night. The annual rainfall can vary from 250 to 750 mm. Veld fires are common during the dry winter months and the plants are well adapted to fires, animal disturbances and drought.

Plectranthus in bushveld regions are markedly succulent (often with succulent roots), or in a few cases annual, which enables them to cope with the seasonal fluctuations of this hot dry climate. Species confined to the flats include *P. caninus*, *P. tetensis*, *P. psammophilus*, *P. tetragonus*, *P. unguentarius* and *P. esculentus*. Those confined to rocky outcrops and dry river valleys include *P. neochilus*, *P. cylindraceus*, *P. spicatus*, *P. amboinicus*, *P. mutabilis*, *P. hereroensis*, *P. unguentarius*, *P. dinteri*, *P. candelabriformis* and *P. verticillatus*. There are a number of species with restricted distributions, found only in certain habitats. *P. psammophilus* occurs only on the very sandy Makatini Flats in northern KwaZulu-Natal. *P. venteri* and *P. porcatus* are confined to the bushveld complex of heavy metal igneous rocks in Limpopo Province, while *P. xerophilus* is restricted to the dolomite outcrops.

Typical bushveld trees include the mopane (*Colophospermum mopane*), various thorn trees (*Acacia karroo*, *A. sieberiana* var. *woodii*, *A. robusta*), bush-willow (*Combretum apiculatum*), *C. molle*, naboom (*Euphorbia ingens*), red-leaved rock fig (*Ficus ingens*), *Kirkia acuminata*, *K. wilmsii*, baobab (*Adansonia digitata*), hairy-leaved rock-fig (*F. glumosa*), tamboti (*Spirostachys africana*), wild pear (*Dombeya rotundifolia*), the moepel (*Mimusops zeyheri*) and wild medlar (*Vangueria infausta*). Common smaller species include many succulents such as *Aloe marlothii*, *A. greatheadii*, *A. zebrina*, *Sansevieria pearsonii*, *Kleinia fulgens* and the savannah bushveld orchid, *Eulophia petersii*.

3

4

The succulent-rich thicket so characteristic of Africa is found in the southern parts of the Eastern Cape, especially in the dry river valleys. Here the characteristic plants are *Euphorbia* spp., spekboom (*Portulacaria afra*), aloes (*Aloe* spp.), ox-tongue (*Gasteria bicolor*), mother-in-law's tongue (*Sansevieria aethiopicum*), pig's ear (*Cotyledon orbiculata*), *Crassula capitella* subsp. *thyrsiflora* and vlakte vygie (*Delosperma ecklonis*). The seven *Plectranthus* associated with dry thickets are *P. madagascariensis*, *P. verticillatus*, *P. spicatus*, *P. cylindraceus*, *P. pentheri*, *P. strigosus* and *P. mzimvubuensis*. The latter is restricted to dry shale cliffs in the inland Mzimvubu River valley, its succulent root tubers enabling it to cope with this arid habitat.

Survival strategies in a hostile winter-dry environment

Unlike trees, whose sturdiness enables them to cope with environmental pressures, smaller species in the forest are easily trampled and, with their shallow roots, are vulnerable to drought and competition. They cannot depend entirely on seed dispersal as trees can. Many have

1 Magwa Falls, habitat of *Plectranthus hilliardiae* subsp. *australe*

2 Afrotemperate forest in the northern Transkei (Eastern Cape), habitat of *Plectranthus ciliatus* and many other forest-dwelling *Plectranthus* species

3 Rocky mountain grassland north-east of Lydenburg (Mpumalanga) in the Spekboomrivier valley near Kliprots, the grassland habitat of *Plectranthus hadiensis*. *P. neochilus* and *P. verticillatus* are confined to the rocky wooded valley below

4 Thicket vegetation near Hankey, habitat of *Plectranthus madagascariensis*

developed an ability to reproduce vegetatively as an extra means of dispersal. The succulent *Plectranthus* are good examples. Their succulence helps them survive when the upper soil layer dries out, but after a disturbance, it also prevents broken stems from dessicating before they have a chance to root. The higher-rainfall forest-dwelling *Plectranthus* are also well adapted to cope with competition from trees and trampling by forest animals. Their semi-succulent nature enables them to remain in a semi-desiccated state for some time, even if all the ground water has been depleted. Trampling by bushbuck, bushpig and porcupine pushes them in the ground, enabling them to sprout. This is one of the reasons why *Plectranthus* thrives in the garden where so much disturbance abounds. The shallow roots make use of the smallest amount of water that penetrates the upper layer, well away from aggressive tree roots lower down.

Chameleon nature

The vegetative parts of the *Plectranthus verticillatus* group (*P. lucidus*, *P. purpuratus*, *P. strigosus*, *P. verticillatus*) can literally change their form to suit their environment, so much so that the plants can appear to be unrelated. In shade *P. purpuratus* subsp. *purpuratus* will grow large, less succulent leaves to compensate for the low light. Under drought or sun stress, or if space is constricted, the same plant will become very compact. Its leaves will be much reduced, markedly succulent and densely arranged along the stem in a cylindrical form like some of the crassulas, eventually overlapping at the ends of the branches. They can also turn purple, which is caused by the presence of anthocyanins, a chemical that blocks out excessive sunlight. We often see this effect in succulent plants too during the dry sunny season. In cultivation these forms will express themselves according to their growing conditions.

Mat-forming and trailing growth

The ability of the mat-forming and trailing *P. madagascariensis* and *P. verticillatus* to root wherever their branches touch ground enables them to cover vast areas, ensuring their survival. Despite heavy disturbance by elephants, buffalo and large leopard tortoises in the thickets, these plants remain common. This vegetative spread ensures more effective survival, though the plants are also able to reproduce by gene exchange.

Succulence and branch drop

Plectranthus spicatus and *P. cylindraceus* are succulent-leaved shrublets from dry river valleys. If the mother plant is destroyed, the broken branches lying on the ground are so succulent that they often survive until the rains arrive, when they will root. Other species that will root when branches drop include *P. neochilus*, *P. madagascariensis* and *P. strigosus*. *P. grallatus* often forms stem tubers in the leaf axils which drop with disturbances, and will root when conditions are favourable.

The best-known example of branch drop is *P. tetensis*. This mat-forming succulent grows in dry mopane veld and other bushveld regions, and is in fact an annual or weak perennial. Towards the end of the growing season, short branches develop in the leaf axils, and without any outside intervention, these drop spontaneously to the ground where some will root during the first spring rains, while the mother plant dies back. This is an example of a 'vegetative annual'. The plant can reproduce sexually as well, but its very effective vegetative strategy seems to have developed at the expense of flowering and seed production, which tend to be underdeveloped.

Chemical defence

Plectranthus belongs to the Lamiaceae, a family famous for the large number of plants with useful aromatic chemicals in them. Many, including *Plectranthus*, have tiny glandular hairs containing phenolic compounds that deter grazing insects. The glandular hairs secrete etheric oils so that touching a plant releases a strong aroma. The use of *P. neochilus* for its pungent smell as an insect repellent is well known. The colonists also used to hang bundles of *P. fruticosus*, the fly spurflower, from the ceiling to deter flies. *P. venteri* is also known as a good insect deterrent.

Unreachable habitat

The bonsai spurflower, *P. ernstii*, and the cliff spurflower, *P. purpuratus*, grow on sheer rocky ledges out of reach of most predators. *P. ernstii* has articulated, succulent stems and very compact growth, while *P. purpuratus* has very succulent leaves compactly arranged on a virtual pseudostem. Both arrangements bring about an increase in succulence and enable the plants to cope with a dry cliff-face habitat. The compact leaf arrangement (almost a pseudostem) also means that the stomata are less exposed, so less moisture escapes.

Survival underground

Fire and frost are common in grassland and savannah. Species such as *P. pentheri* have developed tuberous roots which resprout after a fire. *P. grallatus* and *P. dolichopodus* occur at high altitudes in the Drakensberg where frost can be severe. If the top growth is frosted, they simply drop their aerial parts, sprouting from underground tubers or stems during spring like a dahlia.

Pollination and seed dispersal

Pollination

The diverse floral architecture of *Plectranthus*, especially the shape and length of the tube, reflects diverse pollination strategies. *Plectranthus* are insect-pollinated, and although sunbirds have been noticed on *P. ecklonii*, they are not the main pollinator. Most *Plectranthus* have short-tubed flowers which attract insects such as the honey-bee (*Apis mellifera*), butterflies, bumblebees and bombyliid flies, but there are a few long-tubed species as well which are only found only in South Africa. Christina Potgieter from the University of KwaZulu-Natal has conducted research examining the pollination of eight *Plectranthus* species of varying corolla tube lengths in the KwaZulu-Natal region. She found a correlation between floral tube length and proboscis length, and one can assume that they have co-evolved. While the short-tubed species were pollinated chiefly by the croncerid flies, tabanid flies and anthophorid bees, the two long-tubed species in her study (*P. ambiguus* and *P. hilliardiae* which both have tubes 20–30 mm long) were mainly pollinated by nemestrinid flies with probosces of similar length (Potgieter et al 1999).

Plectranthus tend to have tall flowers easily seen by their pollinators. As a safeguard against self-pollination, the pollen-bearing anthers ripen before the style does. Only once all the pollen is shed do the anthers curve back, and the style, the female part, extends beyond the corolla tube ready to receive pollen from the next visiting pollen-laden insect.

The main flowering of *Plectranthus* takes place during late summer and early autumn (March–April) at the end of the growing season. Rain is usually abundant and the plants are at their peak, with enough water and energy reserves for flowering and seed development. The inflorescence dries during late autumn and winter when dispersal takes place, so that the seed is ready to germinate during the spring or summer rains.

Seed dispersal

Plectranthus seeds are small smooth nutlets 1-2 mm in diameter, usually four per calyx. The calyces are arranged in whorls on extended spikes or panicles and usually enlarge to twice their size after fertilization. They persist for some time, but eventually detach. In many species the calyx has projecting teeth which undoubtedly have a dispersal function.

Annual and short-lived species such as *P. hereroensis* from rocky bushveld regions expend much energy in seed production. They tend to be very floriferous, producing plenty of seed which is shaken out and dispersed locally. The solitary fast-growing forest species such as *P. fruticosus* and *P. ecklonii* also produce large amounts of seed which are dispersed when the tall stiff fruiting branches are shaken either by the wind or a passing animal. If the extended teeth hook onto any creature, the agitation will cause the nutlets to be ejected further, though this will never be far from the parent plant. It is in the best interests of these species to retain their seed in the safest possible habitat, so dispersal is almost always local. Bushveld species such as *P. neochilus* and *P. caninus* in the subgenus *Caleolanthus*, on the other hand, often have calyces which become very brittle when dry. These detach easily, and adhere to the coats of passing animals which disperse them widely. These calyces are also densely hairy at the mouth which prevents the seeds from falling out.

Calyces containing seed can also be dispersed by wind. *P. spicatus* and *P. cylindraceus*, which grow on rocky outcrops of dry river valleys in bushveld and thicket, have extended inflorescences easily disturbed by wind, and small circinate (coiled) calyces which retain their nutlets for some time after they become detached. *P. mirabilis* has flattish seeds which are clearly dispersed by wind from their tall inflorescences.

The cliff-dwelling *P. purpuratus* has a persistent and relatively short inflorescence. The nutlets drop close to the parent, so remaining on the cliff to which it is adapted. Other cliff-dwellers such as *P. ernstii* and *P. dolomiticus* also disperse their seed locally.

Species with protracted flowering such as *P. zuluensis* and *P. saccatus* often have a short inflorescence, and tend not to flower as freely as those with short and dense flowering in early autumn (such as *P. fruticosus* and *P. ecklonii*). Extending the flowering period over a longer period does, however, ensure seed dispersal throughout the summer. These plants tend to be longer lived.

The dispersal of section *Coleoides* (*P. madagascariensis*, *P. grandidentatus*, *P. dinteri*) is mainly by animals. Their calyces are easily detached, and stick easily to the coats of grazing animals which carry them far causing a thicket plant such as *P. madagascariensis* to become widespread.

1

2

1 The succulent root-like tubers of *Plectranthus esculentus* enable the plant to cope with drought. If the tubers are disturbed by man or animal, any remaining fragment will simply resprout

2 *P. woodii* has a tuberous base from which the plant can resprout after fire or frost damage; it also helps it cope with dry winter conditions

7 The genus *Plectranthus* in Southern Africa

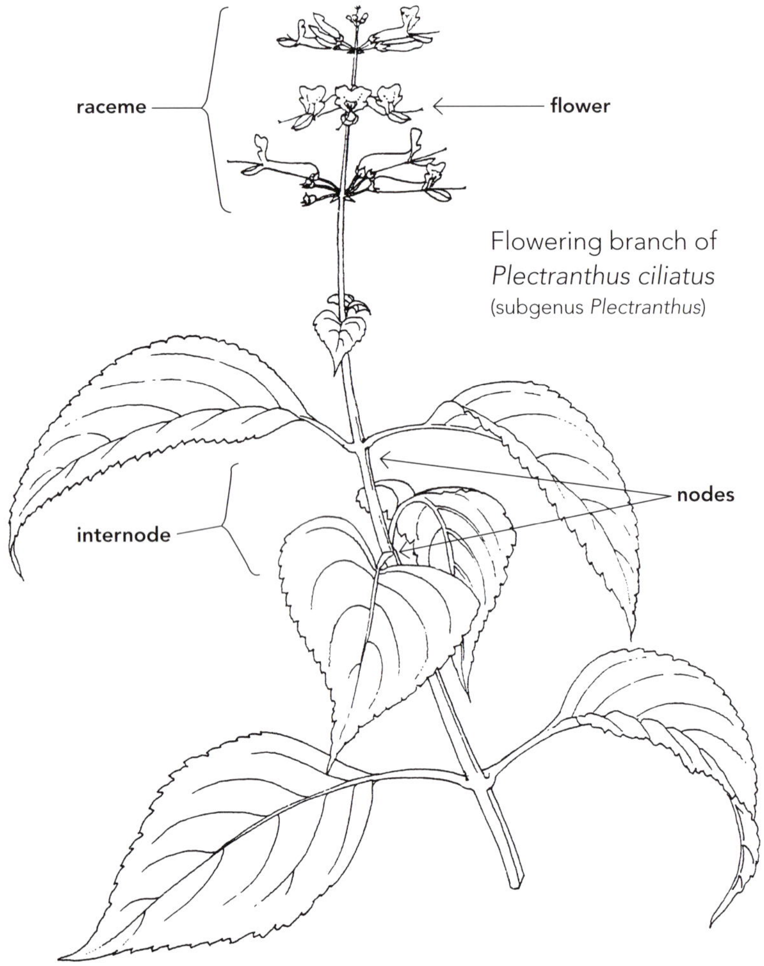

Flowering branch of *Plectranthus ciliatus* (subgenus *Plectranthus*)

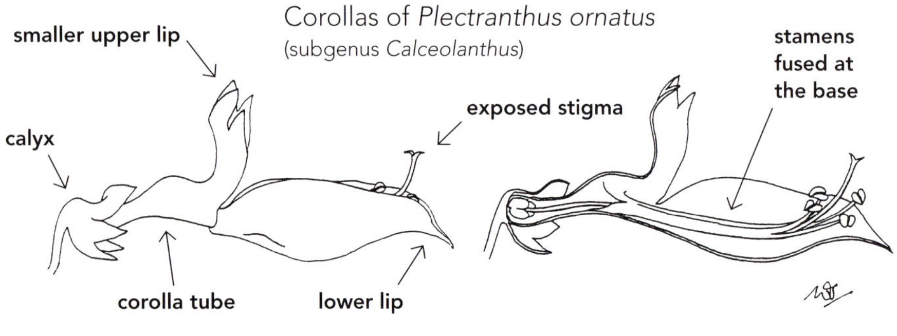

Corollas of *Plectranthus ornatus* (subgenus *Calceolanthus*)

The Lamiaceae

The genus *Plectranthus* belongs to the large cosmopolitan family of the Lamiaceae (Labiatae), which is commonest in the Mediterranean, and comprises roughly 200 genera with about 3200 species (or about 220 genera with 5600 species, depending on which classification one follows). The name Labiatae was used as far back as 1694 and is derived from the two-lipped corolla found in many of its members. The modern family name is derived from *Lamium*, the deadnettle. Other familiar members are the culinary herbs used for their aromatic oils such as sage, mint, lavender, thyme and rosemary.

Considerable variety exists in the family and habitats range from bogs to desert. The members are mostly herbs or shrubs (rarely annuals), occasionally trees (such as *Hyptis arborea* to 18 m), with square stems and leaves in opposite pairs, sometimes whorled, without stipules. Leaves are usually simple and hairy with epidermal glandular hairs, which exude the ethereal oils that give many species an aromatic odour. This feature has been economically exploited in the culinary and cosmetic industries.

The flowers (with bracts) occur mainly in axillary cymes, forming verticillasters on a raceme. The individual flowers are bilaterally symmetrical, bisexual (rarely unisexual) and characteristically two-lipped. The calyx is tubular (and circumscissile at the base in *Aeollanthus*), often enlarging after flowering. The ovary is superior and comprises four dry 1-seeded nutlets.

1

2

This characteristic is the most useful in identifying members of the *Lamiaceae*.

In South Africa and Namibia the family is represented by 36 genera including approximately 274 species, with its main centres of diversity to the east of the Drakensberg escarpment. The best represented genus is *Plectranthus* (53 spp. plus 2 that have naturalized) followed by *Stachys* (41 spp.), *Hemizygia* (28 spp.), and *Salvia* (23 spp.).

The genus *Plectranthus*

The genus *Plectranthus* is distinguished from other members of the Lamiaceae by a number of characteristics. Firstly, the succulent, semi-succulent or softly herbaceous stems root rapidly when detached. Secondly, the two-lipped corolla has exerted stamens (which are directed downwards) attached to its throat. Thirdly, the bracts are smaller than the leaves, and the 5-toothed calyx (with teeth of more or less equal length) enlarges after fertilization. The flowers are carried in 3- to many-flowered cymes, forming whorls or verticillasters on panicles or spikes.

Plectranthus can be confused with *Solenostemon* or *Thorncroftia*, which are also succulent or semi-succulent to softly herbaceous. *Thorncroftia* is at once distinguished by its 4-lobed flowers (the two side lobes drooping from each side) which are solitary in the axils of the leafy bracts. *Solenostemon* is very closely related to *Plectranthus* but separated by its lower calyx teeth which are fused, forming an extended oblong 2-toothed lip.

The classification of plants is not always easy, and the lines are often blurred by the presence of intermediate forms. *Plectranthus* is no exception. It is a very diverse group in which the boundaries are not always clear-cut, and many species seem to grade into others. A number of closely related genera such as *Coleus* and *Burnatastrum* have temporarily been placed under its umbrella, but if a narrower view is adopted, all the subgenera in *Plectranthus* would become independent genera, leaving only 38 species (the present section *Plectranthus*) in the genus. Modern DNA studies may well support this view. Botanists will always differ in their approach to the subject, however, which is not to say that one is wrong and the other right, but that they each follow a different pathway to the same solution.

The genus *Plectranthus* at present includes about 350 species mainly occurring in Africa, but also in Madagascar, India, Japan, Malaysia and Australia (see p. 57). In South Africa the largest concentration of species is found in the coastal north-eastern part of the Eastern Cape (the former Transkei) and southern KwaZulu-Natal (36 species). This area has been described by Van Wyk and Smith as the Pondo centre of plant endemism, and two sites – Port St Johns and the Oribi Gorge – are particularly rich in species. The *P. hadiensis* and *P. madagascariensis* complexes are poorly resolved resulting in identification problems.

1 Himba mothers and daughters in Kaokoveld. The roots of *Plectranthus unguentarius* are used as a deodorant ingredient in their red ochre body lotion

2 A sangoma collecting the roots of *Plectranthus grallatus* in an afromontane forest

Subdivisions of *Plectranthus*

The South African and Namibian *Plectranthus* have been divided by Dr Codd (1975) into six subgenera.

1. Subgenus *Plectranthus* L'Hérit. Characteristics include flowers which are decussate in sessile or shortly pedicelled cymes forming verticils. The calyx is two-lipped with a broad upper lip and a lower lip comprised of four convergent pointed lobes. The subgenus is further divided into two sections.

a) Section *Coleoides* (10 spp.) with 3–15 flowers per cyme and bracts which are shed early during inflorescence development. The species are *P. aliciae*, *P. dinteri*, *P. grandidentatus*, *P. hadiensis*, *P. madagascariensis*, *P. mutabilis*, *P. psammophilus*, *P. ramosior*, *P. venteri* and *P. woodii*.

Plate 5
***Rabdosiella calycina* (formerly *Plectranthus calycinus*) is closely related to *Plectranthus*. It has leaf-like bracts, leaves often in whorls of three, an expanded corolla which is deflexed near its base, and flowers arranged in pedunculate dichasia. It is a grassland shrub to 1 m with a perennial base that sprouts after a fire**

b) Section *Plectranthus* (28 spp.) with 1–3 flowers per cyme and persistent bracts. (There are two exceptions: *P. dolichopodus* and *P. rehmannii* both have more than 3 flowers per cyme.) The species are *P. ambiguus*, *P. ciliatus*, *P. dolichopodus*, *P. dolomiticus*, *P. ecklonii*, *P. elegantulus*, *P. ernstii*, *P. fruticosus*, *P. grallatus*, *P. hilliardiae*, *P. laxiflorus*, *P. lucidus*, *P. malvinus*, *P. mzimvubuensis*, *P. oertendahlii*, *P. oribiensis*, *P. petiolaris*, *P. praetermissus*, *P. purpuratus*, *P. reflexus*, *P. rehmannii*, *P. rubropunctatus*, *P. saccatus*, *P. swynnertonii*, *P. strigosus*, *P. stylesii*, *P. verticillatus* and *P. zuluensis*.

2. Subgenus *Calceolanthus* Codd (4 indigenous, 2 introduced spp.). Characteristics include the hairy mouth of the calyx and the basally fused stamens. South African species are *P. caninus*, *P. neochilus*, *P. pentheri*, *P. tetensis* (introduced spp. are *P. barbatus*, *P. ornatus*).

3. Subgenus *Coleus* (Lour.) Codd. (1 sp., *P. amboinicus*). With staminal filaments which are fused at the base. The upper lip of the calyx is oblong. *P. amboinicus* is the only South African representative.

4. Subgenus *Burnatastrum* (Briq.) Codd. (5 indigenous spp.). With subequally 5-toothed circinate (coiled) calyces, and flowers arranged in dense scorpioid dichasia (short coiled cymes). South African and Namibian representatives are *P. candelabriformis*, *P. mirabilis*, *P. hereroensis*, *P. spicatus* and *P. cylindraceus*.

5. Subgenus *Xerophilus* Codd. (3 indigenous spp.) *P. porcatus*, *P. unguentarius* and *P. xerophilus* are the sole members of this subgenus, characterized by its glomerate verticils and corollas with hooded upper lips.

6. Subgenus *Nodiflorus* Codd. (2 indigenous spp.). With single or paired flowers in terminal pseudoracemes (reduced from cymes, flowers yellow). South African species are *P. esculentus* and *P. tetragonus*.

Key to the South African *Plectranthus* species

FLOWERS IN PSEUDORACEMES

Subgenus *Nodiflorus*
Flowers yellow
2 species

pseudoraceme

yellow flower

Key to subgenus *Nodiflorus*

1a. Annuals, erect . . . . *P. tetragonus*
1b. Trailing deciduous perennials, tuberous roots . . . . *P. esculentus*

Subgenus *Xerophilus*

Upper lip of flower narrow, hooded, obscurely 4-lobed

3 species

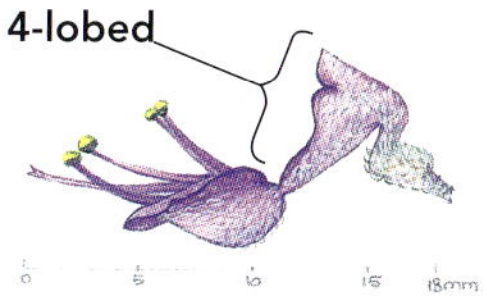

Key to subgenus *Xerophilus*

1a. Older stems distinctly ribbed, rough to the touch . . . *P. porcatus*

1b. Older stems subterete, not rough to the touch . . . 2

2a. Flowers white (leaves with strong pleasant aroma) . . . *P. unguentarius*

2b. Flowers purplish (slightly aromatic) . . . *P. xerophilus*

Subgenus *Burnatastrum*

Mature calyx equally 5-toothed

5 species

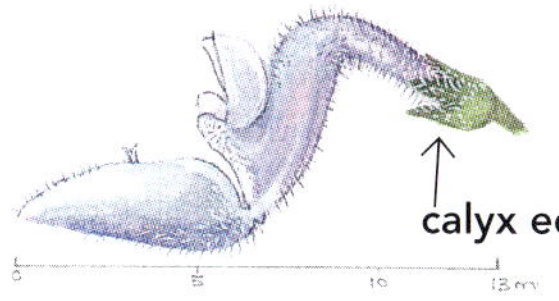

Key to subgenus *Burnatastrum*

1a. Flowers in pedunculate cymes . . . *P. candelabriformis*

1b. Flowers in scorpioid dichasia . . . 2

2a. Leaves small, petiole 5–8 mm, distinctly succulent . . . 3

3a. Inflorescence short, dense, corolla whitish 4–5mm . . . *P. cylindraceus*

3b. Inflorescence in elongated spikes, corolla purple, 7–8 mm . . . *P. spicatus*

2b. Leaves large, petiole 10–70 mm, not distinctly succulent . . . 4

4a. Leaves leathery, whitish below, flowers deep blue . . . *P. mirabilis*

4b. Leaves thin, not white below, flowers bluish . . . *P. hereroensis*

Subgenus *Coleus*

Mature calyx subequally 5-toothed, upper tooth distinctly broader, horizontal

1 species

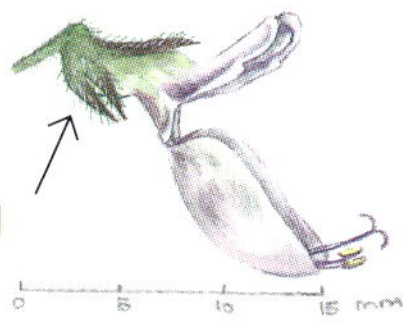

Key to subgenus *Coleus*

Upper tooth of calyx horizontal (pleasantly aromatic) . . . *P. amboinicus*

Subgenus *Calceolanthus*

Mature calyx densely hairy in the throat, not swollen; stamens united at the base

6 species

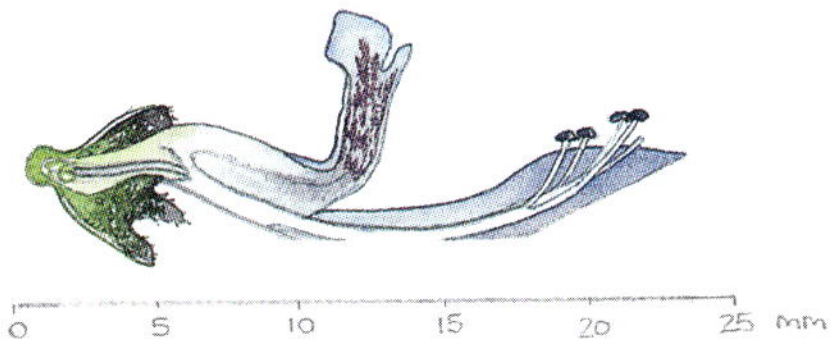

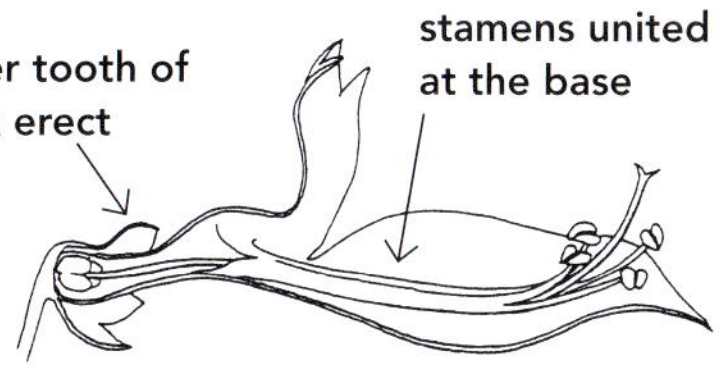

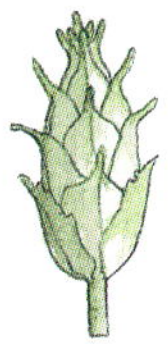

Key to subgenus *Calceolanthus*

1a. Bracts rounded, stems procumbent . . . *P. tetensis*

1b. Bracts acute, forming an imbricate coma . . . 2

2a. Branches decumbent or erect to 50 cm high (distinctly succulent herbs, leaves medium) . . . 3

3a. Roots distinctly tuberous . . . *P. pentheri*

3b. Roots not tuberous . . . 4

4a. Plants annual, corolla less than 10 mm long . . . *P. caninus*

4b. Plants perennial, corolla longer than 10 mm . . . 5

5a. Corolla 10–20 mm long . . . *P. neochilus*

5b. Corolla 20–25 mm long . . . *P. ornatus*

2b. Branches erect, 1.5–3 m, large shrubs (large leaves) . . . *P. barbatus*

Subgenus *Plectranthus*

Mature calyx glabrous in the throat; stamens free to the base

37 species including sections *Coleoides* and *Plectranthus*

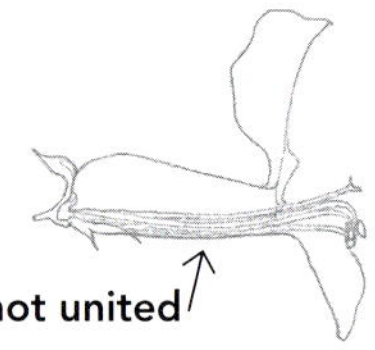

Section *Coleoides*

Bracts drop before flowering; mature calyx distinctly swollen (gibbose) at the base; flowers in up to 15-flowered sessile cymes forming whorls of up to 30 flowers

9 species

Key to subgenus *Plectranthus* (sect. *Coleoides*)

1a. Roots tuberous ... 2
  2a. Leaves ovate, 2–4 cm, flowers mauve (rarely white) ... *P. ramosior*
  2b. Leaves broadly ovate, 4–6 cm, flowers white ... *P. woodii*
1b. Roots not tuberous ... 3
  3a. Corolla 5 mm long, leaves thinly textured ... *P. aliciae*
  3b. Corolla larger than 5 mm ... 4
    4a. Leaves bearing small teeth (4–15 pairs) ... 5
      5a. Branches erect or decumbent ... 6
        6a. Branches 30–60 cm, inflorescence 8–30 cm ... *P. hadiensis* var. *hadiensis*
        6b. Branches 50–150, inflorescence 20–60 cm ... *P. hadiensis* var. *tomentosus*
      5b. Branches procumbent ... 7
        7a. Corolla bluish ... *P. psammophilus*
        7b. Corolla white ... *P. madagascariensis*
    4b. Leaves bearing distinct large teeth (3–6 pairs) ... 8
      8a. Teeth rounded (3–5 pairs) ... *P. mutabilis*
      8b. Teeth triangular ... 9
        9a. Leaves with 3–6 pairs of teeth ... *P. dinteri*
        9b. Leaves with 2 pairs of teeth ... *P. venteri*

Section *Plectranthus*

Bracts persist after flowering; calyx not markedly swollen at the base; flowers in 1–3-flowered sessile cymes forming whorls of up to 6 flowers

28 species

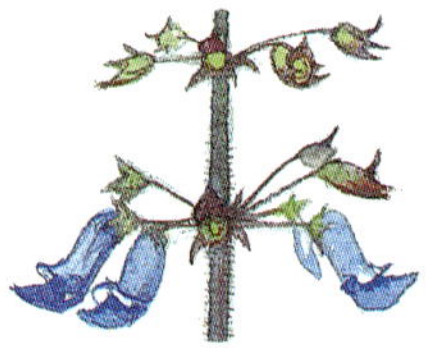

Key to subgenus *Plectranthus* (sect. *Plectranthus*)

1a. Roots tuberous or bearing root tubers ... 2
  2a. Corolla tube curved (declinate) at base ... *P. dolomiticus*
  2b. Corolla tube not curved (declinate) at base, expanding abruptly ... 3
    3a. Stems hollow, flowers mauve ... *P. oribiensis*
    3b. Stems solid ... 4
      4a. Plant erect, dies down to a tuberous base in winter, flowers white ... *P. grallatus*
      4b. Plant a scrambling shrub, flowers mauve ... *P. mzimvubuensis*
1b. Roots not tuberous or bearing root tubers ... 5
  5a. Flowers in 3–8-flowered pedunculate cymes ... *P. rehmannii*
  5b. Flowers in 3-flowered sessile cymes ... 6

6a. Fertile stamens 2 … *P. zuluensis*
6b. Fertile stamens 4 … 7
7a. Corolla abruptly expanding at base (saccate or spurred) … 8
8a. Upper lip of corolla 10–16 mm long … 9
9a. Corolla tube 10–26 mm long … *P. saccatus* subsp. *saccatus*
9b. Corolla tube less than 10 mm long … *P. saccatus* subsp. *pondoensis*
8b. Upper lip of corolla less than 10 mm long … 10
10a. Corolla parallel-sided … 11
11a. Corolla tube longer than 25 mm … *P. hilliardiae*
11b. Corolla tube less than 25 mm long … *P. verticillatus*
10b. Corolla not parallel-sided … 12
12a. Corolla tube 25 mm long … *P. reflexus*
12b. Corolla tube shorter than 20 mm … 13
13a. Corolla distinctly spurred … *P. fruticosus*
13b. Corolla not spurred … 14
14a. Corolla tube about 5 mm long … 15
15a. Erect shrubs … 16
16a. Corolla sky blue (leaf blade thin) … *P. dolichopodus*
16b. Corolla white (leaf blade leathery) … *P. rubropunctatus*
15b. Procumbent to decumbent small plants … 17
17a. Plants (leaves) not succulent … 18
18a. Leaf blade distinctly decurrent on petiole … *P. rubropunctatus*
18b. Leaf blade not distinctly decurrent on petiole … 19
19a. Leaf base cordate, coarsely serrate dentate … *P. swynnertonii*
19b. Leaf base cuneate, finely serrate dentate … *P. elegantulus*
17b. Plants (leaves) succulent, smaller than 3,5 cm … 20
20a. Leaves with rusty or purplish hairs on veins … 21
21a. Leaves green, broadly ovate … *P. strigosus*
21b. Leaves purple below, shiny … *P. purpuratus* subsp. *tongaensis*
20b. Leaves without long, rusty or purplish hairs … 22
22a. Plants compact … 23
23a. Leaves dark purple below … *P. purpuratus* subsp. *purpuratus*
23b. Leaves green, not purple below … *P. purpuratus* subsp. *montanus*
22b. Plants not compact, bearing elongated stems … *P. lucidus*
14b. Corolla tube longer than 5 mm … 24
24a. Basal stems swollen (caudiciform) … *P. ernstii*
24b. Basal stems not swollen … 25
25a. Upper lip of corolla folded backwards … *P. stylesii*
25b. Upper lip of corolla not folded back … 26
26a. Corolla white … 27
27a. Corolla 5–8 mm, leaves ovate … *P. ciliatus*
27b. Corolla 12–13 mm, leaves orbicular … *P. oertendahlii*
26b. Corolla mauve or pink … 28
28a. Leaves semi-succulent, leathery … *P. malvinus*
28b. Leaves thin, not succulent … *P. praetermissus*
7b. Corolla tube not abruptly expanding at base, parallel-sided or gradually expanding … 29
29a. Corolla tube straight … 30
30a. Corolla tube parallel-sided, 2–2.5 cm long … *P. ambiguus*
30b. Corolla tube expanding, 1.2–1.5 cm long … *P. ecklonii*
29b. Corolla tube curved (declinate) … 31
31a. flowers violet-purple … *P. petiolaris*
31b. flowers white … *P. laxiflorus*

Plate 6

Plectranthus aliciae

Origtin: Mtamvuna River,
KwaZulu-Natal (364/89)

8 The species in Southern Africa

Plectranthus aliciae

Plectranthus aliciae (Codd) van Jaarsv. & Edwards in Bothalia 27 (1): 5, (1997). Type: Transkei, near Kentani, Pegler 909 (PRE, holo.).

Name: After Alice Pegler who first collected the species.
Common names: Kentani spurflower, Kentani spoorsalie
Synonym: *Plectranthus madagascariensis* var. *aliciae*

Description: Decumbent, semi-succulent herb, 200–400 mm tall. *Leaves* 25–40 × 22–40 mm, broadly ovate, sparingly strigose, glands red; base truncate, apex rounded to acute; margin with 3–4 pairs of rounded teeth. *Inflorescence* a terminal raceme, 70–130 mm long, occasionally with a pair of side branches, cymes 3–6-flowered. *Corolla* 5–14 mm long, white to light mauve; tube widening to the throat. *Calyx* 3 mm, elongating to 5 mm after flowering. *Nutlets* 1 mm long, light to dark brown.

Flowering time: Autumn (March to May with a peak in April). Occasionally also during summer.

Diagnostic features: *Plectranthus aliciae* is characterized by its leaves bearing only 2–4 pairs of rounded teeth. It is related to *P. madagascariensis* which differs in having 5–7 pairs of teeth and a trailing habit. In his revision of *Plectranthus* Codd (1975) states that var. *aliciae* was known from only two gatherings made by Alice Pegler and perhaps further collections would verify that this taxon is worthy of specific status. It is now known to be locally commonly along the northern part of the Eastern Cape coast. By virtue of its non-trailing stature, very small corolla and membranous leaves with 2–4 pairs of rounded teeth it was raised to specific status.

Distribution: From the Eastern Cape to southern KwaZulu-Natal.

Habitat: It occurs in the shade of subtropical coastal forests where the annual summer rainfall is 600–1250 mm and frost is absent.

History: The species was discovered by Alice Pegler (1861–1929), a teacher and naturalist from Kentani, Transkei (Eastern Cape). It was introduced to Kirstenbosch Botanic Gardens in 1977 (813/77). Mrs Pegler is also commemorated in *Aloe peglerae*.

Cultivation: Easily raised from cuttings or seed, suitable for subtropical forest gardens. Can be planted in shaded rockeries but the white-flowered plants are rather insignificant. Thrives as a pot plant.

Plectranthus ambiguus

Plectranthus ambiguus (H. Bol.) Codd in Bothalia 8: 159 (1964); 11, 4: 429 (1975); F. S. A. 28, 4: 170 (1985). Type: Cape Province.–'near Grahamstown', MacOwan s.n. (BOL, holo.; SAM).

Name: Latin, '*ambiguus*' = doubtful or uncertain, referring to its original placement in another genus.
Common names: pincushion spurflower, speldekussing spoorsalie
Synonyms: *Orthosiphon ambiguus*; *Plectranthus coloratus*; *P. dregei*

Description: Herbaceous perennial 300–700 (1200) mm. *Stems* decumbent to procumbent (rarely erect), rooting at the nodes, pubescent. *Leaves*: blade, 35–100 × 25–90 mm, ovate, slightly rugose, pubescent to glabrescent; base cuneate and somewhat decurrent on petiole; apex acute; margin crenate with 6–12 pairs of teeth; often with long purplish hairs on petioles and veins below; petiole 10–70 mm long. *Inflorescence* a congested terminal raceme or panicle 50–180 mm long. *Calyx* 5 mm enlarging to 8 mm after flowering. *Corolla* bluish-purple to violet; tube straight, 20–30 mm long, 1 mm deep and parallel-sided. *Nutlets* 1.5 mm, black or dark brown.

Flowering time: Autumn (March to May).

Diagnostic features: Related to *Plectranthus ecklonii*, *P. hilliardiae* and *P. reflexus*, all of which have straight corolla tubes. *P. ambiguus* can however be distinguished by its long (20–30 mm) narrow parallel-sided corolla tube 1 mm wide. *P. ecklonii* is an erect shrub with a short expanding corolla tube of 12–15 mm in length, and *P. reflexus* has a corolla saccate for half its length. *P. hilliardiae* differs in having larger flowers 40–45 × 4–5 mm.

Plectranthus ambiguus **grown as a groundcover in the shade of** ***Olinia emarginata***

Distribution: This species occurs from the Eastern Cape to northern KwaZulu-Natal. Considerable geographic variation occurs; in the south plants tend to be smaller and glabrescent.

Habitat: A fairly common forest species confined to coastal subtropical and afromontane forest. It is usually found along forest fringes, in frost-free regions where annual rainfall is 600–1500 mm.

History: *P. ambiguus* was first collected in 1832 by Johann Drège between the Mgazana and Mzimvubu Rivers in the Transkei. It was described as *P. coloratus* by Bentham and Meyer in 1837, but this name had already been given to another *Plectranthus* species. It was first introduced to cultivation at the Durban Botanical Gardens in 1895 by John Medley Wood, and in 1915 Frank Stayner, former curator of the Karoo National Botanical Garden, brought it to Kirstenbosch (4385/15). Dr Bolus described it as an *Orthosiphon* but it was transferred to *Plectranthus* by Dr Codd in 1964. The genus *Orthosiphon* is distinguished by its truly woody members in which the upper calyx tooth forms narrow wings on the calyx tube and the staminal filaments are fused to the corolla well within the corolla tube.

Cultivation: *P. ambiguus* forms an attractive groundcover and could be planted in light shade on the southern or eastern margin of a building or group of shrubs. It is best suited for subtropical forest gardens but is perfectly happy in coastal districts such as Cape Town and Port Elizabeth. It is easily propagated by both cuttings and seed, which germinates in about 21 days. The plants survive light frost. It can be grown in a hanging basket or container, but should then be replanted annually. A number of cultivars have been grown at Kirstenbosch.

'Karkloof'. A diminutive procumbent form, with leaves 25 mm long, and the corolla 24 mm long. The stems have short white hairs and the flowering panicles are 100–120 mm long. The main flowering is at the end of March (1301/82 Kirstenbosch).

Plate 7

Plectranthus ambiguus

Origin: Oribi Gorge, KwaZulu-Natal (797/77)

(*Plectranthus ambiguus* **cont.**)

'Nahoon'. An almost glabrous form, with leaves 50–60 mm long. Peak flowering time is at the end of March (812/77 Kirstenbosch).

'Oribi'. This plant grows to 100–300 mm. The leaves are large (70–110 mm), and are markedly decurrent on the leaf-stalk. The corolla tube is ±30 mm in length. This free-flowering cultivar is resistant to rust. Its flowering peaks at the end of April (797/77 Kirstenbosch).

'Thesiger'. The plant is almost erect (400–600 mm), flowers are shorter than in the typical *P. ambiguus*, and the main flush is in the later half of April (459/89 Kirstenbosch).

'Umque'. The leaves of this cultivar are suffused with purple on the lower surface. It has an erect habit, and flowers at the end of March. It was introduced by Peter Law, former curator of the Natal National Botanical Gardens at Pietermaritzburg (345/76 Kirstenbosch).

Plectranthus amboinicus

Plectranthus amboinicus (Lour.) Spreng. in Syst. Veg. 2: 690 (1825); Codd in Bothalia 11: 388 (1975); in F. S. A. 28, 4: 147 (1985). Type: Amboina, Moluccas, Kerr s.n.(BM).

Name: From Amboinia in the Moluccas, where the plant was first described.
Common names: French thyme, country borage, Indian mint, soup mint, krui-spoorsalie
Synonyms: *Coleus amboinicus*; *C. amboinicus* var. *violaceus*; *C. aromaticus*; *C. crassifolius*; *Majana amboinica*; *Plectranthus aromaticus*

Description: Decumbent distinctly and pleasantly aromatic succulent herb to 150–500 mm, occasionally scandent. *Stems* obscurely 4-angled. *Leaves* 15–50 × 10–40 mm, broadly ovate, pubescent; glands brown to colourless; margin dentate (7–10 pairs of teeth); base truncate; apex obtuse or acute. *Inflorescence* racemose, 100–500 mm long, flowers in 4–10-flowered dichasia. *Corolla* 7–9 mm long, white or mauve, slightly deflexed, expanding to throat. *Calyx* 2–3 mm elongating to 5–6 mm long after flowering. *Nutlets* light to dark brown, 1 mm across.

Flowering time: Late summer to autumn (February to May with a peak at the end of March).

Diagnostic features: *Plectranthus amboinicus* can be confused with *P. cylindraceus*, but the stamens of the former are united at the base, the leaves are basally truncate, the margin has 7–10 pairs of teeth and the verticillasters are laxly arranged. *P. cylindraceus* has leaves with cuneate bases and margins with 3–5 pairs of teeth and densely subspicate inflorescences.

Distribution: Widespread, occurring from Kenya to northern KwaZulu-Natal. It was widely used as a herb (Codd 1975) and has been recorded from Indonesia (Rheede 1689), but its origin is thought to be Africa (Launert 1968, Codd 1975). Plants are common on Inhaca Island off the Mozambique coast and were probably spread by early traders Codd (1975). See also p. 14.

Habitat: Grows in savannah and succulent thicket vegetation on rocky outcrops and cliff-faces in pockets of sandy soil. It is often associated with dry river valleys. Annual rainfall in these habitats varies between 600 and 1000 mm and frost is light or absent. At Mkuze plants were found growing on a northern rhiolite slope both in full sun and partial shade. Associated species included *Plectranthus cylindraceus*, *Euphorbia tirucalli*, *Sarcostemma viminale*, *Kalanchoe sexangularis*, *Portulacaria sp.*, *Aloe spicata*, *Senecio fulgens* and *Commiphora schimperi*.

Uses: Used medicinally in Ceylon, especially in the treatment of cattle diseases. It was traditionally hung from oxcarts (Trimen 1895). It is still a popular herb and sells under the following names: Soup Mint, French Thyme, Spanish Thyme, Country Borage (Launert 1969). It is cultivated in many tropical and subtropical countries. The Jamaicans use it to flavour fish dishes (Launert 1968), and the leaves are also eaten with bread and butter.

History: It was first introduced to Kirstenbosch from cuttings donated by Ian Garland (446/89).

Cultivation: *P. amboinicus* is easily propagated from cuttings and looks effective in a rockery. It is, however, best suited for dry bushveld and succulent thicket gardens, and thrives in well-drained soil under subtropical conditions. The stems are brittle but layer easily. Unfortunately, in humid climates plants are prone to mildew. Some forms are more resistant.

'White Fragrance'. A mildew-resistant variegated white-flowered form that is easily grown.

Plate 8

Plectranthus amboinicus

Origin: From Ian Garland who introduced it from Zululand (KwaZulu-Natal) (446/89)

Plectranthus barbatus

Plectranthus barbatus Andr. in Bot. Rep. t. 594; Codd in Bothalia 11: 394 (1975); in F. S. A. 28, 4: 151(1985). Type: t. 594 in Bot. Rep.; plate drawn from a cultivated plant from Ethiopia introduced into England by Lord Valentia.

Name: '*barbatus*' = bearded, referring to the bearded calyx.

Common names: Abyssinian coleus, bearded spurflower, baard spoorsalie

Synonyms: *Plectranthus forskohlaei*; *Coleus forskohlii*; *C. barbatus*

Description: Erect, branched shrub to 1–3.5 m. *Leaves* 90–150 × 55–100 mm, ovate to broadly elliptical, tomentose; margin crenate with 5–6 pairs of teeth; base cuneate; apex obtuse. *Inflorescence* erect, racemose, 200–450 mm long; bracts large, imbricate and deciduous; rachis viscid, and reddish-purple. *Flowers* in 6–8-flowered verticils. *Corolla* 25–28 mm long, blue-purple. *Calyx* 6 mm long, enlarging to 8–9 mm after flowering. *Nutlets* dark brown, 1.5 mm in diameter.

Flowering time: Summer, but with a peak in autumn.

Diagnostic features: Its size, large leaves and sticky inflorescence distinguish *P. barbatus* from other South African species.

***Plectranthus barbatus* 'Vicki' makes a large shrub with attractive variegated leaves**

Distribution: According to Codd (1985) *P. barbatus* is native to north-eastern Africa and has been recorded from Tanzania, Sudan and Ethiopia. It is naturalized in South Africa and has an invasive status.

Habitat: High-altitude forest margins in its native habitat but it has naturalized, usually growing in waste places in many subtropical areas.

History: The species was introduced into Europe in 1805 by Lord Valentia of Arley from seed he collected in Ethiopia.

Cultivation: *P. barbatus* makes an attractive large shrub with dark blue-purple flowers, popular with gardeners, and is commonly grown where frost is not severe. Plants should be well composted and can placed in full sun or light shade. Because of its invasive status (category 3), it may no longer be sold by nurseries.

'Vicki'. A well-known cultivar with variegated leaves, named after Vicki Thomas.

Plate 9

Plectranthus barbatus

Origin: Ex hort.(117/80)

Plectranthus candelabriformis

Plectranthus candelabriformis Launert in Mitt. Bot. München 7:300 (1968). Type: Namibia, 16 km E. of Runtu, Merxmüller & Giess 1912 (M., holo.).

Name: '*candelabriformis*' = shaped like a candelabra, referring to the inflorescence.
Common names: candelabra spurflower, kandelaar-spoorsalie

Description: Erect branched pleasantly aromatic annual shrub to 1 m. *Branches* ascending, square, striate, sparingly pubescent with long multicellular hairs. *Leaves* ovate, 60–150 × 35–80 mm, sparingly hairy on both surfaces; margin crenate-dentate (15–20 pairs of teeth), the lower surface with orange gland dots; petiole 23–60 mm long; base rounded to almost heart-shaped, apex acute to subacuminate. *Inflorescence* a terminal branched candelabriform panicle to 400 mm long. *Bracts* sessile to shortly petiolate, ovate lanceolate to narrowly elliptic, 5–12 × 5 mm long, persistent beyond flowering, axillary branches. *Flowers* in 3-flowered cymes on slender peduncles to 20 mm long; pedicels 5–10 mm long. *Calyx* 3–5 mm long enlarging to 9 mm (in fruit), densely covered with orange gland dots; teeth subequal, linear-lanceolate up to 4.5 mm long. *Corolla* violet 6.5–7.5 mm, tube slightly kneed at the base, to 3.5 mm, 1 mm in diameter slightly expanding to throat; upper lip erect to 3.5–4 mm long, emarginate with 2 lateral, auriculate lobes; lower lip boat-shaped, to 3.5 mm long. *Stamens* 3 mm long, free at the base and enclosed in the lower lip. *Nutlets* grey, mottled, oblong, 1.8 × 0.8 mm.

Plate 10
Plectranthus candelabriformis
Origin: East of Rundu, Namibia

Flowering time: Summer and autumn (late February to April).

Diagnostic features: *P. candelabriformis* is characterized by its candelabra-shaped inflorescence, its stalked 3-flowered cymes of violet flowers, and its subequal calyx teeth, which is not a typical feature of *Plectranthus*. Another unique feature is the mottled seed.

Distribution: From northern Namibia to Zambia and Tanzania.

Habitat: On the margin of moist seasonal pans and savannah floodplains, usually in semi-shade of shrubs and small trees, in sandy soil. Rain falls mainly in summer and ranges from 700–1000 mm per annum.

History: It was collected near Rundu in northern Namibia by the visiting German botanist Hermann Merxmüller (1920–) and Willie Giess(1910–). Edmund Launert named it *P. candelabriformis* in 1968.

Cultivation: Although this annual with small violet flowers is not particularly attractive, it can be sown in spring to flower in summer in a bushveld garden.

Plectranthus caninus

Plectranthus caninus Roth in Nov. Pl. Spec. 279 (1821); Codd in Bothalia 11: 390 (1975); in F. S. A. 28, 4: 150 (1985). Type: India, Heyne s.n. (Herb. Wallich, K).

Name: From the Latin '*canis*' = dog, meaning inferior.
Common names: dog spurflower, hond-spoorsalie
Synonyms: *Coleus caninus*; *C. flavovirens*; *C. omahekense*; *C. spicatus*

Description: Erect, semi-succulent annual or rarely biennial herb to 200–400 mm. *Leaves* 30–55 × 25–35 mm, elliptic to oblanceolate, slightly pubescent, glands orange-red; margin crenate-dentate (4–7 pairs of teeth); base cuneate; apex obtuse to acute. *Inflorescence* racemose, 25–90 mm long, occasionally with a pair of side branches; bracts large, imbricate, deciduous. *Flowers* in 6–8-flowered verticils. *Corolla* 8–10 mm long, purple, geniculate and expanding to throat. *Calyx* ±2–2.5 mm enlarging to 6 mm after flowering. *Nutlets* dark brown, 1 mm long.

Flowering time: Autumn.

Diagnostic features: *Plectranthus caninus* is a variable species with the material from Limpopo Province having larger flowers. It is closely related to *P. neochilus* but the latter is a decumbent perennial, usually with bigger flowers and whorls more widely spaced. The verticillasters of *P. caninus* are usually closely arranged, covering the rachis. The plant illustrated is not typical; its larger flowers lean more towards *P. neochilus.*

Distribution: Widespread in southern and eastern Africa and India. In Africa it has been recorded from Ethiopia to northern Namibia, North-West and Limpopo Provinces.

Habitat: Savannah, in full sun or semi-shade. It occurs in predominantly summer-rainfall areas where annual precipitation ranges from 400 to 700 mm, and frost is light or absent.

History: *P. caninus* was first described in 1821 by Roth from plants collected by Benjamin Heyne in India. It was introduced to Kirstenbosch by Mrs Mauve from the farm Mosdene in the Limpopo Province (77/81).

Cultivation: It can be grown in warm dry bushveld gardens in full sun or light shade. It is not very attractive and is prone to fungal attack. Plants are easily grown from seed or cuttings.

Plate 11
Plectranthus caninus
Origin: The Springbok Flats north of Pretoria (Gauteng)

Plate 12

Plectranthus ciliatus

Origin: Kirstenbosch (ex hort.) (1262/82)

Plectranthus ciliatus

Plectranthus ciliatus E. Mey. ex Benth. in E. Mey., Comm. 227 (1837); Codd in Bothalia 11, 4: 414 (1975); in F. S.A. 28, 4: 161 (1985). Type: Transkei – 'Omsamvubo' (Mzimvubu River) Drège (K., ex Herb. Benth. No. 4777, lecto.; MO; P; S).

Name: Latin, '*cilia*' = fine hairs. The name refers to the fringe of hairs on the leaf margin, almost like an eyelash.
Common names: eyelash spurflower, wimper-spoorsalie
Synonym: *P. natalensis*

Description: Procumbent to decumbent herb, 200–300 mm high; stems often purple, with purple hairs. *Leaves* 60–200 × 20–65 mm, broadly elliptic to broadly ovate, sparingly strigose with colourless glands, lower surface often purple; base cuneate; apex acute; margin with 9–16 pairs of shallow teeth. *Inflorescence* terminal, racemose 200–280 mm long, often with a pair of side branches, verticillasters 4–8-flowered. *Calyx* 5 mm enlarging to 9 mm after flowering. *Corolla* 8–15 mm long, white or light purplish-pink, saccate at the base and narrowing slightly to the throat. *Nutlets* 1.5 mm, dark brown.

Flowering time: Late summer to winter (mid-February to June, with a peak in April).

Diagnostic features: This species is related to *P. fruticosus* but is distinguished by its decumbent nature and white corolla which is not spurred at the base. *P. fruticosus* is always a shrub and its corolla is either pink or more commonly blue and markedly spurred.

Distribution: From Knysna in the Western Cape to Mpumalanga in the north. It is common in the forests of KwaZulu-Natal and Swaziland.

Habitat: Restricted to moist well-wooded afromontane and subtropical forest, in frost-free, mainly summer rainfall areas where the annual precipitation is 800–1625 mm. It grows on the forest floor in damp places, often in bogs.

History: *Plectranthus ciliatus* was first collected by Drège at the Mzimvubu River in the Transkei. It was introduced to Kirstenbosch by Mrs Carter in 1914 (3789/14), and is still grown extensively as a groundcover. It was taken to Sweden in 1932 by Mrs Karlsberg (Verdoorn 1949), where it is known as the 'Karlsbergia hanging plant'. It is a popular groundcover in America, Europe and Australia.

Cultivation: With its ornamental leaves and white flowers, *P. ciliatus* is a very attractive groundcover, much used today, and best suited to moist subtropical gardens. It also thrives in coastal Cape Town and Port Elizabeth, and will grow on the highveld in protected sites. It is very useful for hanging baskets and containers. Cuttings may be planted *in situ* (about 200 mm apart) during the growing season. Rooting is fast and a closed cover is soon formed. It is best pruned after flowering. If plants become moribund they should be replaced by cuttings.

'Drège'. This commonly cultivated form with dark purple undersides probably comes from the Port St Johns region. It is named in honour of Drège who first collected this species (1268/82 Kirstenbosch).

'Richard'. A small form with hairy leaves and pink flowers, named after Richard Jamieson, a horticulturist specializing in South African plants (1309/82 Kirstenbosch).

***Plectranthus ciliatus* 'Drège' at Kirstenbosch. It makes a vigorous groundcover with delicate white flowers, ideal for suppressing weed growth in shady areas**

Plectranthus crassus

Plectranthus crassus in Gard. Chron., ser. 3, 1904: 21. Type: Malawi (Mahon s.n. K).

Name: '*crassus*' = thick, referring to the succulent stems.

Common names: Mulanje spurflower, Mulanje spoorsalie

Description: Erect, sparingly branched, succulent subshrub to 40 cm high. *Roots* tuberous. *Stems* subterete, succulent and swollen at the bases, young stems green-white. *Leaves* ovate to oblong-ovate, succulent, 40–65 × 40–55 mm, shallow serrate-crenate, with 12–17 pairs of teeth; upper surface green, strigose, lower surface green-white, densely strigose; gland dots red; tip obtuse to acute, base cuneate. *Inflorescence* 18–25 cm long, racemose; verticils subpetiolate, 3–6 flowered; pedicels 2–3 mm long. *Corolla* 20 mm, bluish-purple; tube deflexed, expanding to throat. Fruiting calyx 6–8 mm. *Nutlets* not seen.

Distribution: Only known from Mount Mulanje in Malawi.

Habitat: Grassland among granite rocks, often in association with *P. sanguineus*. Rainfall is mainly during summer and varies between 1500 and 2000 mm a year.

History: *P. crassus* was described in 1904 by N. E. Brown from plants collected by J. Mahon. It was introduced to Kirstenbosch in 1990 (120/91).

Cultivation: This is a striking species with its dark flowers against the whitish undersurface of the leaves. It is best grown as a pot plant or in rockeries in highveld gardens in full sun. It is easily propagated from cuttings and roots rapidly. It is resistant to light frosts and should re-sprout after damage.

Plate 13

Plectranthus crassus

Origin: From the top of Mulanje Mountain, Malawi (120/91)

Plectranthus cylindraceus

Plectranthus cylindraceus Hochst. ex Benth. in DC., Prodr. 12:60 (1848); Codd in Bothalia 11: 385 (1975); in F. S. A. 28, 4: 146 (1985). Type: Ethiopia, Samen, near Gapdia, Schimper 113 (K, holo.; BM; G; P).

Name: '*cylindraceus*' refers to the cylindrical inflorescence.

Common name: camphor spurflower, kanfer spoorsalie

Synonyms: *Germanea cylindracea*; *Plectranthus densiflorus*; *P. marrubioides*; *P. moschosmoides*; *P. spiciformis*; *P. villosus*

Description: Erect to spreading, succulent herb to 200–700 mm. *Stems* 4-angled, decumbent to ascending. *Leaves* 10–50 × 8–40 mm, obovate, pubescent to glabrescent, glands colourless; margin dentate with 3–5 pairs of teeth; base cuneate; apex acute. *Inflorescence* a dense subspicate raceme, 60–350 mm long. *Flowers* in dense decussate (more than 10-flowered) dichasia. *Corolla* 5 mm long, pale mauve, tube straight, expanding to throat. *Calyx* ±2 mm elongating to 3 mm after flowering. *Nutlets* shiny brown, 1.5 mm long.

Flowering time: Autumn into winter (March to June with a peak at the end of April).

Diagnostic features: *Plectranthus cylindraceus* is related to *P. spicatus* but the latter has a lax inflorescence bearing much larger dark violet-purple flowers. The leaves of *P. cylindraceus* are strongly aromatic with a camphor-like scent, while those of *P. spicatus* are only slightly aromatic.

Distribution: From the Bashee River in the Eastern Cape northwards to Ethiopia. It is widespread in tropical southern Africa and also occurs in northern Namibia.

Habitat: Rocky outcrops on sandy substrates in dry woodland in summer-rainfall regions where the annual rainfall is 500–600 mm, and frost is light or absent.

History: It was introduced to Kirstenbosch in 1928 by Mr L. R. Vogts (806/28).

Cultivation: *P. cylindraceus* is an excellent rockery plant suitable for dry bushveld and succulent thicket gardens where frost is not severe. It also thrives as a pot plant. It is easily propagated in spring from seed or cuttings. Young plants should be spaced 200–300 mm apart in light shade or full sun. Germination takes 8–10 days.

'Lebombo' is an attractive form with leaves in a compact, subcylindric body.

'Wylie' is an attractive form from Wylie's Poort in the Northern Province. The leaves are smaller (to 15 mm) and more succulent than in most forms, and are unusual in only having 3 pairs of teeth. It makes a compact, useful groundcover (793/76 Kirstenbosch).

Plate 14

Plectranthus cylindraceus

Origin: The Zoutpansberg, Limpopo Province (793/76)

Plectranthus dinteri

Plate 15

Plectranthus dinteri

Origin: From plants collected by Adam Harrower on the Waterberg in Namibia (530/02)

Plectranthus dinteri Briquet in Bull. Herb. Boissier, ser. 2,3: 1070 (1903). Type: Namibia, Hereroland, Dinter 336 (B).

Name: After Kurt Dinter (1868–1945), German horticulturist, botanist and lover of succulents. From 1894 he was in charge of the famous garden on the Italian Riviera, La Mortola, which had a large collection of South African plants. In 1897 he came to Namibia and was appointed by the German government as botanist to the territory in 1900. He is famous for his botanical exploration in Namibia and is also commemorated in *Aloe dinteri*.

Common names: Waterberg spurflower, Waterberg spoorsalie

Description: Semi-succulent aromatic perennial herb, up to 500 mm. *Branches* square and with glandular hairy surface. *Leaves* 30–90 × 25–70 mm, ovate to broadly ovate; surface glandular-tomentose (candle-like gland-tipped hairs), the lower surface also with red gland dots; petiole to 40 mm long; base truncate; apex acute to rounded; margin coarsely dentate with 3–6 pairs of teeth. *Inflorescence* terminal, racemose up to 250 mm long, occasionally with a pair of side branches; flowers in 3–6-flowered cymes forming 6–12-flowered verticillasters, 10–25 mm apart. *Corolla* 8–10 mm long, mauve to purple, tube deflexed, widening toward the throat. *Calyx* 2–3 mm, enlarging to 4 mm after flowering, base swollen. *Nutlets* 1.2 mm, brown.

Flowering time: Summer to winter (December to June with a peak in April).

Diagnostic features: *Plectranthus dinteri* is closely related to *P. grandidentatus* and *P. madagascariensis*, but is differentiated mainly by its erect growth (procumbent in the others) and its candle-like gland-tipped hairs.

Distribution: *P. dinteri* has a wide but disjunct distribution, from the Waterberg in Namibia to the mountains of the northern Kaokoveld (Omavanda to Otjihipa). It is also found in the western Waterberg of Limpopo Province of South Africa.

Habitat: It is a species of dry savannah, confined to altitudes above 1000 m, almost always growing among rocks (quartzitic sandstone or granite) or on cliffs in well-drained sites, either in full sun or light shade. Plants are subject to seasonal frost and veld fires. Rainfall in these areas ranges from 400 to 600 mm per annum.

History: *P. dinteri* was first collected by Kurt Dinter (see above).

Cultivation: This species is best suited for dry bushveld or highveld gardens. It is easily propagated from cuttings or seed, and grows rapidly, with young plants flowering within a season of sowing. They can be grown in rockeries in full sun, planted 300 mm apart, or in containers. The best time to prune is after flowering.

'Waterberg Maroon'. A form with dark maroon flowers introduced by Kirstenbosch horticulturist Adam Harrower (530/02).

Plectranthus dolichopodus

Plectranthus dolichopodus Briq. in Bull. Herb. Boissier s,r. 2,3: 1069 (1903); Codd in Bothalia 11: 423 (1975); in F. S. A. 28, 4: 167 (1985). Type: Natal, Karkloof, Rehmann 7383 (Z, holo.).

Name: From Greek, '*dolichos*' = long and '*podos*' = foot, referring to the leaf-stalks.
Common names: sky-blue spurflower, hemelblou spoorsalie
Synonym: *Plectranthus cooperi*

Description: Erect herb, 250–700 mm high, main stem with a slight basal swelling. *Leaves* 25–70 × 20–65 mm, ovate to broadly ovate, surface sparingly pubescent; base cuneate; apex acute; margin dentate or crenate, with 8–12 pairs of teeth. *Inflorescence* a terminal raceme or lax panicle, 150–270 mm long. *Corolla* 7–9 (–11) mm long, sky-blue; tube deflexed, gradually enlarging to the throat. *Calyx* 3–4 mm long enlarging to 7 mm after flowering. *Nutlets* 1.5 mm, light brown.

Flowering time: Spring to autumn (October to May with a peak during April).

Diagnostic features: The sky-blue corolla with its diminutive upper lip separates *P. dolichopodus* from other South African members of the genus. The basal swelling of the main stem is also diagnostic among South African species.

Distribution: Widely distributed in afromontane forest along the Drakensberg escarpment and midlands from the Eastern Cape to Limpopo Province. It is also found further north on Mount Mulanje in Malawi.

Habitat: *P. dolichopodus* occurs in moist, cool forest where rain falls mainly in summer and averages between 1000 and 2000 mm per annum.

History: *Plectranthus dolichopodus* was first collected by Dr Anton Rehmann near Pietermaritzburg. It was introduced to Kirstenbosch in 1977 (1080/77).

Cultivation: With its diminutive sky-blue flowers *P. dolichopodus* is an unusual addition to any garden. Plants are best treated as biennials or annuals in cool moist shady forest gardens. It should thrive on the highveld provided it is planted in a protected site. It is easily propagated from cuttings taken during the warmer months. Young plants should be set 450 mm apart in well-composted soil and watered frequently. As a pot plant *P. dolichopodus* is prone to whitefly.

Plate 16

Plectranthus dolichopodus

Origin: Drakensberg foothills, KwaZulu-Natal

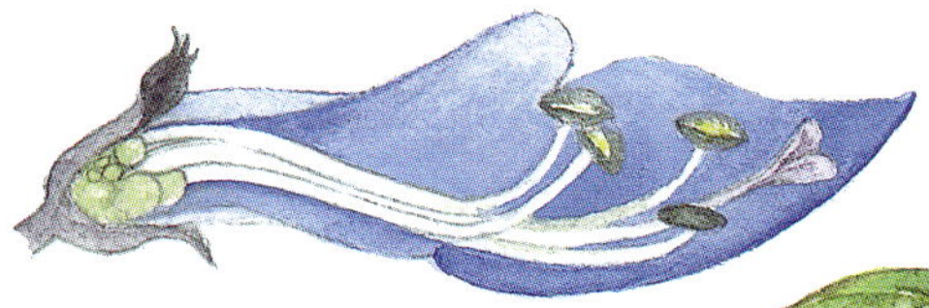

Plectranthus dolomiticus

Plectranthus dolomiticus Codd in Bothalia 15: 142 (1984); Codd in F. S. A. 28, 4: 171 (1985). Type: Transvaal, Van Jaarsv. 7052 (PRE, holo.!).

Name: '*dolomiticus*' refers to the dolomitic substrate on which the plant grows.
Common names: dolomite spurflower, dolomiet spoorsalie

Description: A decumbent, semi-succulent herb, up to 300 mm; rootstock tuberous. *Leaves* 20–30 × 18–30 mm, broadly ovate, surface almost glabrous; base truncate; apex acute or obtuse; petiole up to 35 mm long; margin crenate-dentate with 5–7 pairs of teeth. *Inflorescence* a terminal raceme up to 130 mm long, occasionally with a pair of basal branches. *Corolla* up to 10 mm long, violet-purple, deflexed and widening to the throat. *Calyx* 2.5 mm, enlarging to 6 mm after flowering. *Nutlets* 1.75 mm long, brown.

Flowering time: Autumn (March to April).

Diagnostic features: Recognized by its violet-purple flowers, small semi-succulent, broadly ovate leaves and tuberous roots. This species is closely related to *P. petiolaris*, which has larger leaves and fibrous roots.

Distribution: *P. dolomiticus* was thought to be endemic to the Olifants River valley, near Penge Asbestos Mine in Limpopo Province, but a recent collection from Zimbabwe confirms that it has a wider distribution.

Habitat: In South Africa the species is confined to dolomitic outcrops, mainly cliffs where plants trail from rock crevices in full sun or light shade. In Zimbabwe it has been collected on granite. Rainfall in these habitats is low, averaging 400–500 mm per annum.

History: This species was first collected by the author in 1983 and was introduced to Kirstenbosch in the same year (52/83).

Cultivation: *P. dolomiticus* is easily grown from cuttings. It is drought-resistant and well suited to sun or partial shade in dry bushveld or subtropical gardens where frost is not severe. Young plants should be planted 200 mm apart. It also does well as a pot plant.

Plate 17
Plectranthus dolomiticus
Origin: Penge, Olifantsrivier, Limpopo Province (52/83)

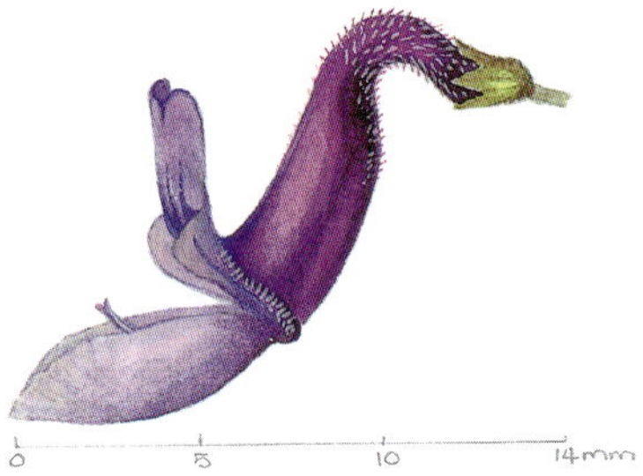

Plectranthus ecklonii

Plectranthus ecklonii Benth. in DC. Prodr. 12: 64 (1848). Codd in Bothalia 11, 4: 428 (1975); in F. S. A. 28, 4: 170(1985). Type: Cape Province – 'Slopes of Katberg', Ecklon s.n. (K, holo.).

Name: After Christian Frederick Ecklon (1795–1868), plant collector and traveller in South Africa.
Common names: Ecklon spurflower, Ecklon spoorsalie

Description: Erect aromatic freely branched shrub to 3 m tall. *Branches* 4-angled, the younger branches strigose, with tufts of purplish hairs at the nodes. *Leaves* 74–190 × 35–115 mm, with petioles up to 50 mm long, ovate to elliptical, slightly rugose and sparingly pubescent, the lower surface with reddish-brown glands; base cuneate; apex acute; margin ciliate, shallowly crenate-dentate bearing up to 25 pairs of teeth. *Inflorescence* a terminal panicle bearing 2–4 pairs of branches, 250–370 mm long. *Bracts* ovate, to 8 mm long and soon deciduous. *Flowers* in sessile 3-flowered cymes, forming 6-flowered verticillasters, the latter up to 15 mm apart. *Pedicels* up to 15 mm long. *Calyx* 5 mm, purple-tinged, elongating to 15 mm after flowering; upper lip erect, ovate, 3 mm long, the lower lip 4-toothed. *Corolla* 15–24 mm long, bluish-purple, rarely pink or white; tube straight, about 1.5 mm broad at the base, dilating to 3.5 mm at the throat; upper lip bilobed, erect, to 6 mm long; the lower lip boat-shaped to 5 mm, and horizontal. *Nutlets* 2 mm, black or dark brown.

Flowering time: Autumn (March to May with a peak in April).

Diagnostic features: A number of *Plectranthus* species display long, narrowly linear corollas including *P. hilliardiae*, *P. reflexus*, and *P. ambiguus*. *P. ecklonii* however can be distinguished from these by its comparatively short dilating corolla, 14–25 mm long, and by the stature of mature plants. *P. ambiguus* has a long (20–30 mm) narrow, parallel-sided corolla tube 1 mm wide. *P. reflexus*, which is also an erect shrub, has a tube which is saccate at the base for half the perianth length. *P. hilliardiae* has a broad parallel-sided corolla ±40 × 4–5 mm. Another diagnostic feature is the leaf sap which stains a yellow-orange colour; rubbed onto the hand it will produce that colour within a minute.

Distribution: *P. ecklonii* is found from Somerset East in the Eastern Cape northwards to Barberton in Mpumalanga. It is common in KwaZulu-Natal coastal forests.

Habitat: It appears to be confined to moist coastal subtropical and mild afromontane forest and forest margins. It usually forms dense localized stands in moist areas with an annual rainfall of 1000–1750 mm.

History: It was first collected in 1813 by the naturalist William Burchell in the Eastern Cape, according to Dr Codd (1983). When Bentham described the plant in 1848, he named it after Ecklon who had collected it on the Katberg. Bentham was apparently unaware of William Burchell's original collection. At the end of the nineteenth century John Medley Wood introduced it to the Durban Botanical Gardens. Frank Bradlow (1965) believes that Ecklon may have sent seed of this species to Baron Von Ludwig in Cape Town. It was certainly grown at Kirstenbosch in 1914, shortly after the gardens were established, and was the first plectranthus to be introduced there (1790/14).

Cultivation: *P. ecklonii* is one of the most attractive in the genus and a deservedly popular garden plant, more grown than any other in South Africa. It thrives in partial shade where frost is not severe, and in cultivation gets to

The white *Plectranthus ecklonii* 'Tommy' at Kirstenbosch. The pink cultivar 'Erma' can be seen in the background

(***Plectranthus ecklonii* cont.**)

1–3 m. It is best suited to shady forest and subtropical gardens, but also does well in coastal districts of Cape Town and Port Elizabeth. It is easily propagated from cuttings which can be planted *in situ*. Young plants grow rapidly and given ample compost and fertilizer will flower the same year. For massing plants should be set out at intervals of 0.5 m. They should be pruned after flowering and at least before spring. Three cultivars are commonly grown.

'Erma'. This pink-flowered form of unknown provenance is named after Erma van Jaarsveld, the author's wife. The leaves and stems are also tinged with pink (216/82 Kirstenbosch).

'Medley Wood'. This tall blue-flowered form is the most commonly grown. It is named after John Medley Wood who introduced the species into cultivation (1306/82 Kirstenbosch).

'Tommy'. This white-flowered cultivar is of unknown provenance. Dr Codd once told the author that a white form of *P. ecklonii* was known from Kentani in the Transkei. This one is named after Tommy Sardien, head gardener of the succulent section at Kirstenbosch, who grew many of the *Plectranthus* species at Kirstenbosch so successfully (1307/82 Kirstenbosch). It is slightly shorter and less vigorous than the others.

Although both 1 'Erma' (pink) and 2 'Tommy' (white) have been in cultivation for at least the past 30 years, their origin in the wild remains unknown. The young stems and leaf stalks of 'Erma' are flushed with pink as well, so the plant can be recognized even without its flowers. The leaves of *P. ecklonii* are unusual in leaving a nicotine-like stain on the skin if they are crushed and rubbed

1

2

Plate 18

Plectranthus ecklonii **'Medley Wood'**

Origin: Kirstenbosch (ex hort.) (1306/82)

(pink) 'Erma'
(white) 'Tommy'

Plectranthus elegantulus

Plectranthus elegantulus Briq. in Bull. Herb. Boiss. Der. 2,3: 1005 (1903). Type: Natal, Karkloof, Rehmann 7368 (Z, holo.; PRE, photo.).

Name: '*elegantulus*' = elegant.
Common names: elegant spurflower, elegante spoorsalie.

Description: Soft decumbent and spreading herb to 600 mm high. *Branches* succulent 4-angled, with short hairs. *Leaves* with petiole to 40 mm long, the blade thinly textured, broadly ovate, to 40 × 50 mm; margin serrated, with 4–8 pairs of teeth; surface sparingly hairy and beset with minute honey-coloured glands on the lower surface; base cuneate; apex acute to obtuse; often tinged purple on the lower surface. *Inflorescence* terminal, simple or with a pair of side branches, racemes to 12 cm long; rachis bearing short hairs; bracts ovate-lanceolate to 4 mm long. *Flowers* in 1–3-flowered cymes, forming up to 6-flowered verticillasters, the latter up to 15 mm apart; pedicels to 6 mm. *Calyx* 4 mm long enlarging to 8 mm when seed matures; the upper lip erect, ovate and up to 3 mm long; the lower lip 4-toothed. *Corolla* white or light mauve with a few darker purple dots, to 10 mm long, base saccate and tapering to the mouth, the tube about 4 mm long; upper lip erect to 5 mm long, bilobed and bearing two small lateral lobes; lower lip boat-shaped to 3 mm long. Stamens free up to 2 mm long. *Nutlets* dark brown, 1.5 mm in diameter.

P. elegantulus is a groundcover from the KwaZulu-Natal midlands. It is similar to *P. ciliatus*, but smaller. It flowers profusely

Flowering time: Autumn (March to May with a peak towards the middle to the end of March).

Diagnostic features: Similar to *P. ciliatus* but more erect, with smaller leaves and fewer hairs.

Distribution: Only known from the KwaZulu-Natal midlands near Karkloof.

Habitat: A groundcover in cool moist afrotemperate forest and forest margins in summer-rainfall areas where annual rainfall is between 1500 and 2000 mm. This plant grows together with *P. ciliatus*.

History: *P. elegantulus* was described in 1903 by Briquet from plants collected by Rehmann at Karkloof. It was introduced to Kirstenbosch in 1981 (118/82).

Cultivation: A floriferous plant which thrives in cultivation, it is best grown as a groundcover for shade in subtropical and forest gardens. It is easily propagated from seed or cuttings. Germination takes place within 3 weeks. Plants can be set about 300 mm apart. It does well as a pot plant and is striking in hanging baskets.

Plate 19

Plectranthus elegantulus

Origin: Karkloof, KwaZulu-Natal (118/82)

Plectranthus ernstii

Plectranthus ernstii Codd in Flower. Pl. Afr. 47: 1855 (1982). Type: Natal, Oribi Gorge, Van Jaarsv. 2196. (PRE, holo.!).

Name: After Ernst van Jaarsveld (1953–), horticulturist at the Kirstenbosch National Botanical Gardens.
Common name: bonsai mint

Description: Erect to decumbent succulent herb, up to 250 mm high; stems swollen and variably articulated, up to 50 mm in diameter. *Leaves* ovate to broadly ovate; blade 12–30 × 10–25 mm, the base truncate; apex acute; margin dentate or crenate, with 3–5 pairs of teeth; surface sparingly pubescent or almost glabrous. *Inflorescence* a terminal raceme 50–170 mm long, verticillasters 2 to 6-flowered. *Calyx* 3 mm enlarging to 6 mm after flowering. *Corolla* 5–12 mm long, saccate at the base and gradually narrowing to the throat, light blue with darker markings. *Nutlets* 2 mm long, light brown.

Flowering time: October to May with a peak during April. Also sporadically throughout the year.

Diagnostic features: The articulated succulent stems are the result of anomalous secondary thickening of predominantly parenchymatous tissue, a unique feature amongst South African species. These stems superficially resemble those of the genus *Thorncroftia* which thicken in response to infestation by the weevil, *Apion rectangulum* Wagn. *P. ernstii* is allied to *P. strigosus*, *P. praetermissus* and *P. reflexus* which are florally similar. The species is variable with several local forms.

***Plectranthus ernstii* can be grown in rockeries. It is a small species with whitish to light mauve flowers**

Distribution: From the northern part of the Eastern Cape (Msikaba River) to Oribi Gorge in southern KwaZulu-Natal.

Habitat: *P. ernstii* occurs on exposed to semi-exposed, sheer quartzitic sandstone rockfaces of the Cape Supergroup (Natal Series) in river gorges from the Msikaba to Oribi Gorge. The branches are often covered with lichen and are well camouflaged amongst the sandstone rocks. Associated succulents include *Gasteria croucheri*, *Adromischus cristatus* var. *zeyheri*, *Crassula perforata*, *C. perfoliata*, *Delosperma* sp. and *Aeollanthus parvifolius*. In these areas rain falls in summer and averages between 800–1500 mm per annum. Frost is absent or light.

History: It was first collected in 1977 on the lip of Oribi Gorge by the author and his wife Erma (805/77 Kirstenbosch).

Cultivation: It thrives as a pot plant and can be successfully grown in shallow pans on windowsills. In subtropical areas it will do well in a rockery crevice. The leaves turn rose-pink during autumn with the drop in temperature; the species is frost-sensitive. It thrives in coastal areas such as Cape Town and Port Elizabeth provided it is watered during summer. Plants grow best in full sun or light shade and are easily propagated from soft tip cuttings or from seed, germinating in ±21 days. Prune frequently to maintain a neat 'bonsai' shape. Four forms have been introduced.

'Forster's Foley'. Branches procumbent with dark green leaves and dark blue mauve flowers, discovered by Tony Abbott on the farm Forster's Foley along the Mtamvuna River (431/89 Kirstenbosch).

'Msikaba'. The leaves of this form from the Msikaba River are very succulent and the branches more elongated, tending to droop (790/89 Kirstenbosch).

'Oribi'. The form was found in the Oribi Gorge in southern KwaZulu-Natal. It has decumbent branches with light green leaves and light blue flowers. It is popular as a 'bonsai' subject (805/77 Kirstenbosch).

'Sikuba'. This cultivar has flaccid elongated stems and dark purple vertical striations on the upper lips of the corolla (276/06).

Plate 20

Plectranthus ernstii

Origin: Oribi Gorge, KwaZulu-Natal (805/77)

Plectranthus esculentus

Plectranthus esculentus N.E. Brown Kew Bull. 12 (1894); Codd in Bothalia 11: 377 (1975); in F. S. A. 28. 4:142 (1985). Type: Natal, cult. Botanic Garden, Durban, Medley Wood 3633 (K, holo.).

Name: Latin, '*esculentus*' = edible, referring to its edible roots.
Common names: veld potato, veldartappel
Synonyms: *Englerastrum floribundum*; *Coleus daxo*; *C. esculentus*; *C. floribundus*; *Plectranthus floribundus*; *P. floribundus* var. *longipes*

Description: Trailing, semi-succulent, deciduous herb to 600–1200 mm. *Roots* tuberous. *Leaves* deciduous, 50–80 × 13–25 mm, oblong-elliptic, shortly scabrid, abaxial surface glandular; margin denticulate; base obtuse or cuneate; apex obtuse. *Inflorescence* 50–80 mm long, comprising axillary racemes. *Flowers* pedicellate, solitary. *Corolla* 14–16 mm long, yellow, expanding to the throat. *Calyx* 4–5 mm, elongating to 10 mm after flowering. *Nutlets* 1 mm long.

Flowering time: Spring.

Diagnostic features: *Plectranthus esculentus* is characterized by its tuberous roots, trailing habit and yellow flowers, which appear after the leaves are shed. It is the only spring-flowering species. There is one other yellow-flowered species, *P. tetragonus*, but that is an erect annual with erect stem bristles, glabrescent leaves and smaller flowers.

Distribution: From KwaZulu-Natal to Limpopo Province and into tropical Africa.

Habitat: Sandy, well-drained soils in rocky savannah where annual rainfall is 700–1000 mm, and frost is light or absent.

General: *P. esculentus* is an indigenous food popular for its high starch content. It is cooked like a potato. The Zulu vernacular name is 'umbondwe'.

History: The species was described in 1894 by N. E. Brown from plants collected by Medley Wood and grown in the Durban Botanical Gardens. It was introduced to Kirstenbosch in 1934 by a Mr Hepburn from Nigerian stock (2832/34).

Cultivation: It is best suited to dry sandy bushveld gardens, where, unusually in the genus, it will produce its yellow flowers in winter and spring. Easily propagated from cuttings under warm subtropical conditions; these should be planted 600 mm apart in full sun. This species does not thrive in the winter-rainfall region of the Western Cape.

Plectranthus esculentus **puts out its yellow flowers on bare stems during winter and early spring**

Plate 21

Plectranthus esculentus

Origin: Kirstenbosch (ex hort.)

Plectranthus fruticosus

Plectranthus fruticosus L'Hérit. Stirp. Nov. fasc. 4: 85, t.41 (March 1788); Codd in Bothalia 11, 4: 415 (1975); in F. S. A. 28, 4: 162 (1985). Type: t.41 of L'Hérit., Stirp Nov. fasc. 4 (1788).

Name: Latin, '*fruticosus*', shrubby, referring to its habit.
Common name: muishondblaar
Synonyms: *Germanea urticifolia*; *Plectranthus arthropodus*; *P. behri*; *P. charianthus*; *P. galpinii*; *P. peglerae*; *P. urticifolius*

Description: Erect shrub (rarely decumbent) to 1.5–2 m. *Branches* with purple hairs at the nodes. *Leaves* 90–150 × 70–120 mm, often purple below, broadly elliptic to broadly ovate, sparingly pubescent, glands colourless; base cuneate or truncate; apex acute; margin with 7–27 pairs of shallow teeth. *Inflorescence* a terminal panicle 200–280 mm long, verticillasters 2 to 6-flowered. *Calyx* 5 mm, enlarging to 9 mm after flowering. *Corolla* 10–13 mm long, blue, purple or pink, saccate or distinctly spurred at the base, narrowing to the throat. *Nutlets* 1.5 mm, dark brown.

Flowering time: Autumn (middle of March to May with a peak in April).

Diagnostic features: *P. fruticosus* is easily recognized by the distinct spur at the base of the corolla (hence the name of the genus). *P. ciliatus* is superficially similar but is always a groundcover with white flowers lacking the spur of the corolla. *P. fruticosus* has upright growth with panicles of blue or pink flowers.

The very free-flowering 'Ellaphie' has leaves suffused with purple and a tendency to spread, unlike the normal *P. fruticosus* which has erect growth

Distribution: From Caledon in the Western Cape to Limpopo Province, following the moist afromontane forests.

Habitat: Moist subtropical or afromontane forest where it is locally abundant, usually in forest margins. Soils are well-drained and humus-rich. Rainfall, predominantly in winter in the Western Cape, elsewhere in summer, averages 1000–1750 mm per annum.

History: It was first collected by Carl Thunberg and Francis Masson in the southern Cape in 1774. Masson sent seed to Kew, from where it was widely distributed across Europe. It was the first plectranthus to be grown as an ornamental. In his famous work *Arboretum et Fruticetum Britannicum* (1838) J. C. Loudon (1783–1843) described seeing it in France:

> *Plectranthus fruticosus* is a native of the forests near the Cape of Good Hope, an old inhabitant of our green-houses, and one of the few greenhouse plants that was found in old conservatories in France before the revolution ... [Here] among the old orange trees, pomegranates, olives, and oleanders, which are occasionally found lingering about the few old chateaux that still exist, *Plectranthus fruticosus* may be found sometimes 6 ft or 7 ft high. In an area of a house in Berkeley Street, there were, in 1836, two plants, about 6 ft high, and of proportionate bulk. Mr Bowie, in a communication to the Garden Magazine on raising Australian and Cape shrubs from seeds, and acclimatizing them to Europe, proposes the use of this species in green-houses. He suggests that since it is most susceptible to cold, if properly placed will provide a warning to guard against direct injury to other plants, as it is always the first to suffer.

It was introduced to Kirstenbosch in 1914 by Louisa Bolus (1841/14).

Cultivation: *Plectranthus fruticosus* is an attractive and very popular garden shrub. The purple undersurface of the leaves of many forms is also most decorative. It is at its best in moist forest gardens but also does well in Cape Town and Port Elizabeth. In the highveld it will need protection from frost. It should be cut back after flowering. Five cultivars have been introduced.

'Behr's Pride'. This form from Lusikisiki in the Transkei was discovered in 1931 by Mr Behr and was donated to Kirstenbosch (1252/31). It was initially given specific status by Compton, but Codd (1973) showed

Plate 22

Plectranthus fruticosus 'James'

Origin: South of Barberton (1077/77)

(*Plectranthus fruticosus* **cont.**)

that it is simply a form of *P. fruticosus*. The leaf undersides are dark purple and the panicles are a striking pink.

'Ellaphie'. This is the only form of *P. fruticosus* which is decumbent. Its leaves are dark purple beneath, the flowers are mauve, and the corolla is weakly spurred. It originates from near Port St Johns in the Eastern Cape, and is named after Mrs Ellaphie Ward-Hilhorst, the well-known botanical artist who has illustrated many *Plectranthus* species (113/82 Kirstenbosch).

'James'. A shrub of about 1.5 m with pink panicles. This form of *P. fruticosus* can be planted in full sun. It was discovered by the ecologist James Culverwell in Mpumalanga near the Barberton border growing on exposed granite outcrops. The broadly ovate leaves are semi-succulent and the veins and stems are covered with purple hairs (1077/77 Kirstenbosch).

'Ngoye'. A pink-flowered form with broadly ovate to orbicular leaves (which are occasionally ternate). The young branches are covered with dense dark red-purple hairs. It was introduced into cultivation by Ian Garland, of Twin Streams, Mtunzini (1305/82 Kirstenbosch).

'Pauline'. This form is identical to 'Behr's Pride' but with mauve flowers. It is named after Pauline Fairall, formerly of the Compton Herbarium (447/89 Kirstenbosch).

***Plectranthus fruticosus* 'James' makes a fast-growing shrub which can tolerate full sun in coastal regions**

Plectranthus grallatus

Plectranthus grallatus Briq. in Bull. Herb. Boissier ser. 2,3: 1004 (1903); Codd in Bothalia 11, 4: 418 (1975); in F. S. A. 28, 4: 165 (1985). Type: Transkei, Mount Frere, Schlechter 6415 (Z, holo.; GRA; PRE.).

Name: Latin, '*grallator*' = a stilt-walker, perhaps referring to the leaf hairs.
Common names: Drakensberg spurflower, Drakensberg spoorsalie
Synonyms: *Plectranthus cooperi*; *P. krookii*; *P. natalensis* f. *glandulosa*; *P. praetermissus*; *P. transvaalensis* var. *grandifolia*

Description: Herb with subterranean tuberous stems, spreading vegetatively; aerial stems erect, herbaceous or shrubby, to 200–1500 mm, with purple hairs at the nodes. *Leaves* 65–135 mm × 40–100 mm, broadly ovate, pubescent, glands red; base truncate or cuneate; apex acute; margin with 14–20 pairs of shallow teeth. *Inflorescence* a terminal panicle, 120–300 mm long, verticillasters up to 8-flowered. *Calyx* 4 mm, enlarging to 7 mm after flowering. *Corolla* 14–17 mm long, white speckled with purple, or rarely light purplish-pink, saccate at the base, narrowing to the throat. *Nutlets* 1.5 mm, dark brown.

Flowering time: Midsummer to autumn (December to May with a peak during February).

Diagnostic features: Its perennial tubers and annual stems are distinctive among white-flowered species. It is related to *P. fruticosus* and *P. oribiensis*. *P. fruticosus* is distinguished by its perennial stems which are basally woody and its mauve to pink corolla; *P. oribiensis* has cordate leaves, hollow stems and a hairy mauve corolla.

Distribution: From Cathcart in the Eastern Cape to Limpopo Province.

Habitat: It is found at high altitudes (up to 3000 m) in afromontane forest and forest margins and is also common in drainage lines in montane sourveld. Plants are often locally abundant. Soils are variable and often permanently damp during the growing season. Frost

Plate 23

Plectranthus grallatus

Origin: South-east of Kaapse Hoop (Mpumalanga) (786/76)

occurs in these regions and summer rainfall averages 1000–1750 mm per annum.

History: It was first collected by Rudolph Schlechter at Mount Frere and was introduced to Kirstenbosch in 1976 (786/76).

Cultivation: *P. grallatus* spreads vegetatively by subterranean horizontal tuberous branches. In the wild it often produces aerial tubers in the leaf axils as well. It is dormant during winter when the tubers can be lifted and divided, or stem cuttings can be taken in early spring. Plants thrive in a moist position in light shade, so a cool forest garden suits it, but it also thrives in Johannesburg as well as in coastal districts. It is frost-resistant.

Plectranthus grandidentatus

Plectranthus grandidentatus Gürke in Bull. Herb. Boissier 6:554 (1898); Codd in Bothalia 11: 396 (1975); in F. S. A. 28, 4: 152 (1985). Type: Cape, East Griqualand, Emeyembe Mountain, Tyson sub Herb. Austr. Afr. 1517 (K, lecto.).

Name: Latin, '*grandidentatus*', with large teeth
Common names: big-teeth spurflower, groottand spoorsalie

Description: Procumbent, trailing, soft herbaceous perennial. *Branches* tomentose. *Leaves* 20–70 × 18–75 mm, ovate to broadly ovate, pubescent; petiole 25–80 mm long; base cordate; apex acute; margin dentate with 4–7 pairs of teeth. *Inflorescence* terminal, racemose, up to 225 mm long, occasionally with a pair of basal branches. *Corolla* 7–13 mm long, white or occasionally mauve; tube slightly deflexed, widening to the throat. *Calyx* 3 mm long, enlarging to 5 mm after flowering. *Nutlets* 8 mm, dark brown.

Flowering time: Autumn into winter (March to June with a peak in May).

Diagnostic features: *P. grandidentatus* is characterized by its procumbent tomentose stems and 4–7 pairs of large teeth on the leaf margin, to which the specific epithet refers. It can be confused with *P. madagascariensis* but that is decumbent and has much smaller rounded teeth. Hybrids between the two occasionally occur. Some degree of introgression is also thought to occur with *P. hadiensis* which produces occasional specimens with rather rounded leaf serrations and mauve flowers. Vegetative material may be confused with *P. swynnertonii* but this species has membranous leaves, sometimes with dense purple hairs, and is restricted to moist forest floor habitats.

Distribution: Widespread from Queenstown in the Eastern Cape to the Zoutpansberg in Limpopo Province.

Habitat: The plants occur on the Drakensberg foothills in rocky screes or on dry rocky outcrops in well-drained humus-rich soils. Summer rainfall is 600–1000 mm a year.

History: *P. grandidentatus* was first collected by Anton Rehmann and William Tyson in the Eastern Cape and was introduced to Kirstenbosch Gardens in 1922 by H. P. Thomasset (1011/22).

Cultivation: Easily propagated from cuttings, *P. grandidentatus* grows rapidly forming an attractive groundcover rooting at the nodes. It does well in lightly shaded forest and other moist bushveld gardens where frost is not severe. Young plants grow rapidly and given ample compost and fertilizer often flower in the same season. Its large teeth are very decorative.

'Hlobane'. An attractive vigorous groundcover with densely hairy leaves 7 × 5 cm. This cultivar originates from Hlobane, KwaZulu-Natal (709/02 Kirstenbosch).

'Liesl'. With its large graceful leaves (12 × 9 cm) this form makes an excellent groundcover. Named after Liesl van der Walt, horticulturist at Kirstenbosch, who introduced this form from Kings Holt, KwaZulu-Natal (218/01 Kirstenbosch).

Plate 24

Plectranthus grandidentatus

Origin: Kirstenbosch (ex hort.) (1011/22)

Plectranthus hadiensis

Plectranthus hadiensis (Forssk.) Schweinf. ex Spreng., Wein 3. Gart. Zeitung 19: 2(1894). F. S. A. 28, 4: 153(1985). Type: Yemen, Hadiya, Forssk†l (C, holo.).

Name: From Hadiyah in the Yemen, where it was first described.

Common names: hairy spurflower, harige spoorsalie, lehlati, ilozane, imbozisa (Zulu)

Synonyms: var. *hadiensis*: *Ocimum hadiense*; *Plectranthus forskohlaei*; *P. pachyphyllus*

var. *tomentosus*: *P. tomentosus*; *P. zatarhendi* var. *tomentosus*

Description: Herbaceous or shrubby perennial to 1.5 m, from a persistent rootstock; branches spreading, tomentose. *Leaves* 35–100 × 25–125 mm, ovate to rounded; semi-succulent, woolly-tomentose, glands red; base cuneate to subcordate; apex acute to rounded; margin dentate or crenate with 4–14 pairs of teeth. *Inflorescence* terminal, racemose up to 500 mm long, occasionally with a pair of side branches; flowers in 4–15-flowered verticillasters, 10–30 mm apart. *Corolla* 8–15 mm long, mauve or occasionally white; tube slightly deflexed, widening to the throat. *Calyx* 2–3 mm, enlarging to 5–6 mm after flowering. *Nutlets* 8 mm, dark brown.

Flowering time: Autumn into winter (March to June with a peak in May).

Diagnostic features: *Plectranthus hadiensis* is variable but is characterized by its shrubby nature, usually densely tomentose leaves, shallow teeth and resprouting rootstock. In rare instances specimens with coarse teeth and pale flowers are found; these are due to introgression with *P. grandidentatus*. It is closely related to *P. madagascariensis* and *P. grandidentatus*, but the latter has large teeth and white flowers while *P. madagascariensis* has much smaller leaves with a trailing habit, crenate leaf margins and white flowers.

Distribution: *P. hadiensis* has a very wide distribution from KwaZulu-Natal to the Arabian Peninsula. The var. *tomentosus* occurs in KwaZulu-Natal and along the Mpumalanga Drakensberg. It is a large erect shrub with subrotund leaves and occasionally tuberous roots. The var. *hadiensis* is pleasantly aromatic and occurs from the KwaZulu-Natal coast to Gauteng.

Habitat: Dry savannah, usually in well-drained humus-rich soils among rocks, where the annual rainfall is 700–1000 mm.

History: Pehr Forsskål discovered this species in Hadiyah in the Yemen and named it *Ocimum hadiensis* in his 1775 flora of Egypt and Arabia. It was the first *Plectranthus* to be described (though the genus had not yet been established). It was only transferred to *Plectranthus* in 1894 by the botanists Schweinfurth and Sprenger. *P. hadiensis* var. *tomentosus* was introduced to Kirstenbosch in 1917 by Carnegie (1091/17).

Cultivation: This species is easily propagated from cuttings and grows rapidly in moist subtropical bushveld gardens, either in light shade or full sun. *P. hadiensis* var. *tomentosus* is especially attractive, and can be grown where frost is not severe. It can become untidy and needs to be pruned. There are three cultivars.

'Carnegie'. Makes a free-flowering shrub to about 1 m. The flowers are light purple (132/31 Kirstenbosch).

'Mamba Valley'. A groundcover with densely hairy and succulent leaves. This cultivar originates from near Durban (902/88 Kirstenbosch).

'Shongweni'. Has attractive wedge-shaped leaves with only 2–3 teeth (728/02).

Plate 25

Plectranthus hadiensis

Origin: Kirstenbosch (ex hort.)

Plate 26

Plectranthus hereroensis **'Kaoko'**

Origin: Okakora Spring, Kaokoveld, northern Nambia (30/02)

Plectranthus hereroensis

Plectranthus hereroensis Engl. in Bot. Jb. 10: 267 (1888); Codd in Bothalia 11: 382 (1975); Codd in F. S. A. 28, 4: 145 (1985).
Type: Namibia, Hereroland, Kaiser Wilhelmsberg near Okahandja, Marloth 1350 (B, holo.).

Name: '*hereroensis*', after Hereroland in Namibia where the plant was first collected.
Common names: Herero spurflower, Herero spoorsalie
Synonyms: *Coleus myrianthus*; *Neomuellera damarensis*; *Plectranthus aurifer*; *P. matabelensis*; *P. myrianthus*; *P. otaviensis*

Description: Erect, prominently branched shrub, 1–2.3 m tall, annual or biennial. *Leaves* 40–70 × 35–70 mm, ovate to ovate-triangular, membranous, glabrescent to sparingly pubescent, glands red; margin crenate-dentate (15–20 pairs of teeth); base truncate, apex acute. *Inflorescence* paniculate 100–250 mm long; flowers in decussate dichasia. *Corolla* 6–11 mm long, blue-purple, occasionally white; tube deflexed, expanding to throat. *Calyx* 2.5 mm, elongating to 5–7 mm after flowering. *Nutlets* shiny, brown, 1 mm.

Flowering time: Autumn into winter (March to June with a peak in May).

Diagnostic features: Within the subgenus *Burnatastrum*, *P. hereroensis* is characterized by its membranous broad leaves, short-lived (occasionally annual) habit and geniculate corolla tube.

Distribution: Widely distributed in the eastern half of South Africa, from near East London in the south to Limpopo Province in the north, as well as Botswana, northern Namibia and Zimbabwe.

Habitat: Occurs in hilly terrain in sandy, dry savannah where annual summer rainfall is 400–750 mm and frost is light or absent.

History: *Plectranthus hereroensis* was first collected by Hermann Wilhelm Rudolf Marloth (1855–1931) in 1886. It was named by Heinrich Engler (1844–1930) in 1888 from plants collected by Marloth on the Wilhelmsberg near Okahandja. It was first introduced to Kirstenbosch by F. G. Carnegie in 1932 (3060/32, Kirstenbosch).

Cultivation: An attractive floriferous shrub suitable for gardens where frost is not severe. It is easily propagated from seed, germinating in 12–21 days, or cuttings in spring. Plants grow rapidly reaching 1–1.5 m in a single season. Young plants are best set 0.7 to 1 m apart in light shade or full sun. It will do well in highveld and bushveld gardens and can be treated like a annual.

'Kaoko'. A very free-flowering attractive dark purple annual originating from Kaokoveld in northern Namibia.

'Malmokkie'. An attractive white-flowered form of unknown provenance.

'Matoppos'. A shrub to 1.75 m with dark purple flowers in April. From southern Zimbabwe.

'Witpoortjie'. An attractive form with light-blue flowers from Gauteng (486/77 Kirstenbosch). See plate on the contents page.

Plectranthus hereroensis **from Okakora Spring in Kaokoveld, northern Namibia, where it is widespread. This is a very showy form. The eastern forms tend to have smaller flowers and are more like the South African plants**

Plectranthus hilliardiae

Plectranthus hilliardiae Codd in Bothalia 11: 282 (1974); in F. S. A. 28, 4: 169 (1985). Type: Cape Province – 'near Mtamvuna River', Hilliard & Burtt 6767 (PRE, holo.; NU!).

Name: After Olive Mary Hilliard (1925–), formerly Professor of Botany at the University of Natal.

Common names: Pondo spurflower, Pondo spoorsalie.

Description: Erect to decumbent, semi-succulent, herbaceous perennial to 200–300 mm. *Branches* hairy, with long purple hairs at the nodes. *Leaves* 25–100 × 15–55 mm, obovate to elliptical, semi-succulent; strigose, glands colourless; base cuneate; petiole 15–35 mm long; apex acute to rounded; margin ciliate, shallowly crenate-dentate in the upper half. *Inflorescence* terminal, racemose 80–180 mm, occasionally with a pair of side branches, verticillasters up to 6-flowered. *Calyx* 4 mm long, elongating to 10 mm after flowering. *Corolla* 27–40 mm long; pale purplish blue with purple spotting; tube laterally compressed, 4–6 mm broad, linear; base saccate. *Nutlets* 2 mm, black or dark brown.

Flowering time: Autumn into winter (February to June with a peak in mid-April).

Diagnostic features: The linear corolla tube is distinguished by its size, and also by being parallel-sided and laterally compressed. *P. reflexus* is the closest ally but this species has a tube which broadens towards the base. Two subspecies are recognized. Subsp. *australe* has elliptical leaves usually suffused with dark purple veins on the underside; subsp. *hilliardiae* has obovate leaves without the purplish markings.

Distribution: This fairly recently described species is restricted to river gorges in the northern Eastern Cape as far as the Mtamvuna River, the southern border of KwaZulu-Natal. Subsp. *australe* occurs from Port St Johns northwards to Fraser Gorge, while subsp. *hilliardiae* occurs north of Fraser Gorge.

Habitat: It occurs on deeply shaded river banks within moist subtropical forest, where it is often locally abundant.

History: *Plectranthus hilliardiae* subsp. *hilliardiae* was first collected in 1919 along the Mpahlane River west of the Mzamba by the botanical explorer H. J. Thode (1859-1932), a German who settled in Natal. His specimen lay unnoticed, however, until after Olive Hilliard and Bill Burtt made their discovery in the Mtamvuna gorge. It was introduced to Kirstenbosch in 1979 (111/82). Subsp. *australe* was introduced to Kirstenbosch during 1992 (146/92) from plants collected at Magwa in the Eastern Cape.

Cultivation: *P. hilliardiae* is one of the most beautiful species. It is best suited to moist subtropical forest gardens, otherwise it can be grown as a pot plant indoors or in a greenhouse. Although cuttings take readily, this species is not as easily grown as its relatives, requiring high humidity and the right temperatures.

'Magwa' A fine form of *P. hilliardiae* subsp. *australe* with almost hairless leaves that have an attractive shiny surface and dark purple undersides (146/92). It has been used in hybridizing, and was involved in the development of 'Mona Lavender'.

Plectranthus hilliardiae × *P. strigosus* (*P. brevilabrum* Edwards)

Plectranthus brevilabrum has recently been named by Trevor Edwards from the University of KwaZulu-Natal (*Bothalia* 35, 2: 149–52, 2005). It is a beautiful small plant found by David Styles in the Luputana Gorge in the Eastern Cape. Its specific epithet '*brevilabrum*' refers to its short lower lip. (*brevis* = short, and *labrum*=lip).

Plectranthus brevilabrum is regarded by the present author, who has grown the plant from cuttings received from David Styles, as a hybrid between *P. hilliardiae* and *P. strigosus* for three reasons. The first is that only a single clone was found growing among *P. hilliardiae*, even though other parts of the gorge were also thoroughly searched. The second is that the plant seems to be intermediate between *P. hilliardiae* and possibly *P. strigosus* (the lower lip of *P. strigosus* is also short and boat-shaped, and the stems also covered in hairs and bearing red gland dots). Thirdly, the flowers of *P. brevilabrum* in cultivation are often distorted, the anthers are frequently without pollen, and seed is either poorly set or not set at all.

Description: Small decumbent semi-succulent herb to 10 cm high. *Stems* densely covered with purplish hairs. *Leaves* obovate 30–65 ×10–30 mm, dark green and covered with red gland-dots. *Inflorescence* a small panicle to 17 cm tall, bearing 3-flowered cymes. *Corolla* 18 mm, light mauve, the upper lip erect, 4-lobed, the two central lobes much larger and laterally pointed with central magenta nectar spots, the lower lateral lobes much smaller; the lower lip boat-shaped, 4,5 mm long, and dark magenta.

Plate 27

Plectranthus hilliardiae
subsp. *australe*

Origin: Magwa Falls, Eastern Cape (146/92)

Plectranthus laxiflorus

Plectranthus laxiflorus Benth. in E. Mey., Comm. 228 (1837); Codd in Bothalia 11: 434 (1975). Type: Natal, between Mzimkulu and Umkomaas Rivers, Drège 3586 (K, lecto.; P; S.).

Name: Latin, '*laxiflorus*', loose-flowered, referring to the lax arrangement on the inflorescences.
Common names: citronella spurflower, sitrinella spoorsalie, ubebebe
Synonyms: *Germanea laxiflora*; *Plectranthus hylophilus*

Description: Erect or decumbent, soft, herbaceous perennial to 300–600 mm, stems glandular-pubescent. *Leaves* 60–100 × 40–60 mm, broadly ovate-deltoid, rugose, pubescent, glands pink to red; base cordate, petiole 25–80 mm long; apex acute; margin dentate with 16–20 or more pairs of teeth. *Inflorescence* a loose panicle or terminal raceme, up to 300 mm long. *Corolla* 12–14 mm long, white, often with pale mauve-brown stripes on the upper lip; tube deflexed, widening to the throat. *Calyx* 3.5–4 mm, enlarging to 8 mm after flowering. *Nutlets* 1.5 mm, brown.

Flowering time: Autumn into winter (March to June).

Diagnostic features: *Plectranthus laxiflorus* has a distinct citronella-like aroma, unique among the South African plectranthus. It is further distinguished by its deltoid-ovate leaves and white sigmoid corolla.

Distribution: It is widespread east of the escarpment from Humansdorp in the Eastern Cape northwards into tropical Africa.

Habitat: Moist places, along stream-banks on forest margins and in clearings of mist-belt vegetation. Summer rainfall ranges from 700 to 2000 mm per annum.

History: It was first collected in 1832 by Johann Drège, and was introduced to Kirstenbosch in 1920 (849/20) by Carnegie from Pietermaritzburg.

Cultivation: *P. laxiflorus* is easily grown from cuttings. It develops into a shrub, rooting at the nodes, and is well suited to shady frost-free sites in moist subtropical forest gardens. Although it is one of the less attractive species, its fresh citronella scent makes it worth cultivating. It is, however, known to be allergenic, causing severe skin irritation in some individuals. Symptoms usually appear 5 to 10 hours after contact.

Luputana Gorge in the north-eastern part of the Eastern Cape, habitat of *P. laxiflorus* and several other *Plectranthus* species such as *P. hilliardiae*, *P. ambiguus*, *P. ciliatus* and *P. zuluensis*. *P. laxiflorus* tends to prefer moist depressions close to the stream bank

Plate 28

Plectranthus laxiflorus

Origin: Kirstenbosch (ex hort.)

Plectranthus lucidus

Plectranthus lucidus (Bentham) van Jaarsv. & T. J. Edwards (BT 27 (1): 5, 1997). Type: Eastern Cape, Burchell 3924 (K.).

Name: '*lucidus*' = shiny, referring to its leaves
Common names: shiny spurflower, blinkblaar spoorsalie
Synonym: *Plectranthus strigosus* var. *lucidus*

Description: Procumbent succulent herb, rooting at the nodes, glabrous or sparsely pubescent; roots fibrous. *Stems* 4-angled, purplish green, 2.5 mm in diameter, strigose with white multicellular hairs. *Leaves* ovate to broadly ovate, 12–15 × 10–15 mm, serrate to crenate-dentate, with 3 or 4 pairs of teeth, green to purplish; abaxial surface sparsely strigose; veins densely strigose, rubropunctate; base cuneate; apex obtuse to acute. *Raceme* simple, 30–100 mm long, rarely with a pair of side branches; cymes 1-flowered; verticillasters 5–10 mm apart; bracts ovate-lanceolate, 2–3 mm long; pedicel 4–5 mm long. *Calyx* at flowering 3–4 mm long, enlarging to 7 mm, densely strigose, with longer purplish hairs. *Corolla* 9–11 mm long, mauve to rarely white; tube straight, 5 mm long, basally saccate, constricted dorsally; upper lip 5 mm long, 2-lobed; upper lobes 3 mm long, lateral lobes 2 mm long, with long multicellular white hairs; lower lip boat-shaped, margins strongly involute. Nutlets brown to black, 1.5 × 1 mm.

Flowering time: Autumn (March to May).

Diagnostic features: *Plectranthus lucidus* is closely related to *P. verticillatus* and *P. strigosus*, but is distinguished by its 2-flowered verticillasters and short corolla (9–11 mm) with the lower lip margins conspicuously inrolled. In both *P. strigosus* and *P. lucidus* the corolla is constricted at the throat, unlike *P. verticillatus* where the corolla is linear.

Distribution: The Eastern Cape from Bathurst northwards to Port St Johns, always following the coastal sand dunes.

Habitat: Plants form mats on stabilized coastal sand dunes below trees, sharing this habitat with other shade-loving species such as *P. madagascariensis*, *Scadoxus membranaceus* and *Dracaena aletriformis*.

History: *P. lucidus* was first collected by William Burchell near Bathurst and described by Bentham in 1848, but only transferred to *Plectranthus* in 1997. It was introduced to Kirstenbosch in 1974 (181/74).

Cultivation: This resilient mat-former withstands salt spray and is useful for coastal and seafront gardens in warm subtropical areas. It does well in Cape Town and Port Elizabeth provided it is given some moisture during summer. Easily increased from cuttings and fast-growing.

'Albert'. A white-flowered form (181/74 Kirstenbosch).

'Bulolwe'. A mauve form introduced from Second Beach, Port St Johns (114/82 Kirstenbosch).

Plate 29

Plectranthus lucidus **'Bulolwe'**

Origin: Port St Johns, Eastern Cape (114/82)

Plate 30

Plectranthus madagascariensis

Origin: Kirstenbosch (ex hort.)

Plectranthus madagascariensis

Plectranthus madagascariensis (Pers.) Benth. in Lab. 37(1832); Codd in Bothalia 11: 402 (1975); in F. S. A. 28,4: 155 (1985). Type: Cape, East Griqualand, Emeyembe Mountain, Tyson sub Herb. Austr. Afr. 1517 (K, lecto.).

Name: '*madagascariensis*', of Madagascar.
Common names: candle plant, thicket spurflower, ruigte spoorsalie

Description: Procumbent to decumbent semi-succulent perennial, branches tomentose. *Leaves* 15–35 × 12–40 mm; ovate to broadly ovate, glands red, strigose above and tomentose below; base cuneate to truncate; apex acute; margin crenate or crenate-dentate with 3–7 pairs of teeth. *Inflorescence* terminal, racemose, 90–125 mm long, occasionally with a pair of basal branches. *Corolla* 7–18 mm long, white; saccate at the base, tube straight, parallel-sided. *Calyx* 3 mm, elongating to 5 mm after flowering. *Nutlets* 1 mm long, light to dark brown.

Flowering time: Autumn to early winter (February to May with a peak in late March and early April).

Diagnostic features: *Plectranthus madagascariensis* is characterized by its trailing procumbent or decumbent stems and 3–7 pairs of small rounded teeth. It is related to a number of other species. *P. mutabilis* has blue to purple flowers; *P. aliciae* is a decumbent to erect herb with 2–3 large rounded teeth; *P. grandidentatus* has large, coarse teeth; *P. woodii* is alone in having a persistent tuberous rootstock, and *P. psammophilus* has much smaller flowers.

Distribution: Widespread from the Eastern Cape to northern KwaZulu-Natal.

Habitat: It is very common in subtropical thicket, occurring in dry rocky outcrops, thorn thicket and forest margins where annual rainfall is 600–1250 mm. It often grows among succulents such as *Aloe africana*, *A. speciosa*, *Euphorbia ledienii*, *Crassula ovata*, *C. cordata*, *Portulacaria afra*, *Senecio pyramidatus* and *Plumbago auriculata*.

History: It was first collected by Philibert Commerson in Mauritius or Réunion some time before his death in 1775, as well as by Thunberg and Masson in 1774. It was introduced to Kirstenbosch in 1914 (4869/14) by the forester J. M. Sim.

Cultivation: *P. madagascariensis* is easily grown from cuttings, forming a quick and attractive groundcover. Although it thrives in dry succulent thicket, it can grow in light shade in most South African gardens where frost is not too severe. It is a very long-lived, drought-tolerant and 'waterwise' plant. It is sometimes wrongly sold under the name 'Plectranthus coleoides marginatus'.

'Lynne'. This popular cultivar has variegated leaves, and is resistant to light frost. It grows vigorously to 100–200 mm in light shade or sun, rooting at the nodes. Named after Lynne Nichols who illustrated so many *Plectranthus* species in South Africa.

Plectranthus madagascariensis **'Lynne' is a natural mutation with mottled colouring. The leaves can revert to plain green, or even mutate further to pure yellow which will eventually cause the stem to die from lack of chlorophyll. This cultivar thrives in hanging baskets and containers, and is a very self-sustaining hardy plant**

Plectranthus malvinus

Plectranthus malvinus Van Jaarsv. & T. J. Edwards in Bothalia 27 (1): 1–2 ill., 1997. Type: Port St Johns, Van Jaarsv. & Bingham 10522 (NBG).

Name: '*malvinus*' = mauve, referring to its mauve flowers.

Common names: mauve spurflower, pienk spoorsalie

Description: Decumbent, strigose, mat-forming herb to 200 mm and 500 mm across; roots shallow, fibrous. *Stems* 4-angled, purple-green, ± 3 mm in diameter, strigose (white multicellular hairs), punctate; internodes 10–40 mm apart. *Leaves* fleshy, firm, ovate to obovate, 40–90 × 30–50 mm, strigose, serrate with 8–10 pairs of teeth; abaxial surface strigose; veins purple, densely strigose; hairs white; gland dots colourless, sunken; base cuneate; apex acute; petiole 5–10 mm long, purple, densely villose on the adaxial surface, decurrent. *Inflorescence* a raceme or lax panicle, 180–210 mm long; cymes 3-flowered, 10–15 mm apart; bracts ovate-lanceolate, 7 × 2 mm long; pedicels ± 10 mm long. *Calyx* 4 mm long (enlarging to 10 mm), upper lobe ovate, 2 mm long (5 mm in fruit), lower lobes 4, linear, 1.5 mm long (4 mm after flowering). *Corolla* ± 12 mm long, pink (violet group 84c), tube laterally compressed, 6 mm long, saccate at base and 3 mm deep, narrowing to 2 mm at throat; upper lip 7 mm long, 2-lobed; lateral lobes 3 mm long, lower lobe boat-shaped, 5 mm long. *Nutlets* brown to black, ovoid, 1.5 × 1 mm.

Flowering time: Autumn (March to May).

Diagnostic features: This species is related both to *P. strigosus* and *P. ciliatus*. It is distinguished by its firm, succulent, ovate to obovate leaves which are serrate, have densely pilose purple veins and are densely punctate underneath. The latter feature separates it from *P. strigosus*, which has red gland dots. The small flowers are reminiscent of *P. ciliatus*, but are generally smaller and an attractive mauve. *P. ciliatus* usually has white flowers and soft leaves.

Distribution: Endemic to the upper region of the quartzitic sandstone mountains, Mt Sullivan and Mt Thesiger, around Port St Johns in the Eastern Cape.

Habitat: Plants occur in forest margins and often on cliffs. Rainfall averages between 1500–2000 mm per annum and falls mainly from spring to autumn.

History: *P. malvinus* was described by Trevor Edwards and the author in 1997. It has been grown as a groundcover at Kirstenbosch since its collection by Alison Bingham and the author in 1989 under the name *P. ciliatus* 'Bingham' (221/90). (See p. 22 for the story of its discovery.)

Cultivation: *P. malvinus* is best suited to warm subtropical forest gardens, and is easily propagated from cuttings and seed. It is an attractive garden plant, has bright mauve-pink flowers and with its succulent leaves is more drought-tolerant than *P. ciliatus*. It also thrives in hanging baskets and containers.

Plate 31

Plectranthus malvinus

Origin: Mount Sullivan, Port St Johns, Eastern Cape (221/90)

Plectranthus mirabilis

Plectranthus mirabilis (Briq.) Launert in Mitt. Bot. München 7: 299 (1968). Lectotype: Angola, Malange, Mechow 489.

Name: '*mirabilis*', remarkable, referring to its striking appearance.

Common names: Water spurflower, water spoorsalie

Synonyms: *Coleus mirabilis*; *Ascocarydion mirabile*

Plate 32
Plectranthus mirabilis
Origin: Popa Falls, Okavango River, northern Namibia (322/06)

Description: Erect shrub 1–3.5 m tall. Young branches slightly square, later rounded, usually several from the base and sparingly branched; surface grey-tomentulose (short multicellular hairs and orange gland dots) and covered with lenticels in the lower part. *Leaves* 60–120 × 30–60 mm, ovate to ovate-lanceolate with petiole to 2 cm long; upper surface dull green, lower surface white-felted; base obtuse to cuneate, apex acute; margin crenate-dentate (25–30 pairs of teeth). *Inflorescence* erect dense sparingly branched panicle to 300 mm (the rachis densely grey-tomentulose). Flowers in many-flowered dichasia arranged in opposite pairs; bracts broadly ovate, acuminate 9–12 × 7–8 mm, early deciduous. *Flowers* in 6 to 8-flowered verticils. *Corolla* 13–15 mm long, blue; tube narrow and cylindrical and abruptly deflexed, expanding to 4 mm at the throat; upper lip erect, 4 mm long (obscurely 4-lobed). *Stamens* united at the base, enclosed or slightly exerted. *Calyx* 4–5 mm, enlarging to 7–8 mm at maturity. Nutlets obscurely winged.

Flowering time: Autumn.

Diagnostic features: Its tall erect usually unbranched *rounded* stems, with lenticels at their bases, its blue flowers and white undersurfaces distinguish it immediately from any other Southern African plectranthus. Another distinguishing feature is its almost winged light-coloured seed which is wind-dispersed.

Distribution: Southern tropical Africa including northern Namibia, Angola, Malawi, Zambia and Zaire.

Habitat: It is unique among *Plectranthus* for only growing on river banks or in wet areas. At Poppa Falls along the Kavango River in northern Namibia it was observed growing among reeds (*Phragmites australis*), *Cyperus papyrus*, *Morella serrata* and *Salix mucronata*. The distinctive lenticels on the lower stems help it take up oxygen in its waterlogged habitat.

History: Mechow collected his plant at Malange, Angola.

Cultivation: *P. mirabilis* is an attractive large shrub which should be used more widely. It could be a handy marginal pond introduction in bushveld and subtropical gardens. With its grey-white leaves and striking blue flowers it makes an arresting sight when in flower.

Plectranthus mutabilis

Plectranthus mutabilis Codd in Bothalia 11: 404 (1975); in F. S. A. 28, 4 156 (1985). Type: Transvaal, Blouberg, Codd 7953 (PRE, holo.).

Name: Latin '*mutabilis*' = changeable, referring to its variability.
Common names: chameleon spurflower, verkleurmannetjie spoorsalie

Description: Prostrate, semi-succulent herb, branches often hanging from cliffs. *Leaves* 15–50 × 15–50 mm, ovate to broadly ovate, tomentose to glabrescent, glands orange; base truncate; apex acute or obtuse; margin crenate or crenate-dentate with 4–6 pairs of teeth. *Inflorescence* terminal, racemose, 100–250 mm long, occasionally with a pair of side branches. *Verticillasters* 6–14-flowered. *Corolla* 8–15 mm long, purple-blue; tube centrally deflexed, widening to the throat. *Calyx* 2 mm, elongating to 4 mm after flowering. *Nutlets* 1 mm, dark brown.

Flowering time: Late summer into winter (February to June with a peak in April).

Diagnostic features: *Plectranthus mutabilis* is characterized by its prostrate, free-rooting stems and its purple-blue flowers. It is related to *P. madagascariensis*, *P. aliciae*, *P. grandidentatus* and *P. woodii*, all of which have white flowers.

Distribution: Widespread in Limpopo Province, Mpumalanga and Gauteng.

Habitat: Usually confined to river valleys and mountain slopes in dry savannah, often drooping from cliff-faces. Plants grow in well-drained and humus-rich soils in light shade. Annual rainfall is 600–1250 mm.

History: It was discovered in 1937 at Trigaardspoort near Pretoria by Jack Repton, horticulturist working for the Pretoria Municipality, but it was only formally described in 1975. It was introduced to Lowveld Botanic Gardens in 1976 (788/76).

Cultivation: The prostrate habit and drought-resistance of *P. mutabilis* make it an ideal groundcover. Plants are easily propagated from cuttings and grow rapidly. It is best suited to moist to dry bushveld gardens. It will also thrive on the highveld but needs frost protection. Ideal for slopes.

Plate 33
Plectranthus mutabilis
Origin: Sudwalla Caves, Mpumalanga (788/76)

Plate 34

Plectranthus mzimvubuensis

Origin: Along the Mzimvubu River, Transkei, Eastern Cape (531/04)

Plectranthus mzimvubuensis

Plectranthus mzimvubuensis Van Jaarsv. in Bothalia 34, 13–32 (2003). Type: near Ludonga Village, Mzimvubu River, Ecca Group shale cliffs, (-AD), Van Jaarsv., Xaba, Harrower & Zwide 92 (PRE, holo.).

Name: After the Mzimvubu River in the Transkei, E. Cape. The Xhosa name *mzimvubu* means 'the home of the hippo' (*Hippopotamus amphibius*).
Common name: Mzimvubu spurflower

Description: Perennial aromatic shrub to 1 m, 3 m across, scandent and pendent from cliffs. *Roots* fibrous to slightly fleshy, but bearing distinct oblong to rounded grey tubers 25–50 × 14–20 mm. *Stems* herbaceous, semi-succulent, 4-angled, terete in older branches and with a succulent basal caudex, 100 mm across. *Bark* smooth, grey. *Leaves* broadly ovate-deltoid to subrotund, 25–50 × 28–50 mm; apex acuminate, with short drip-tip; base truncate to subcordate; occasionally slightly decurrent on petiole, thin-textured; surface covered with slightly sunken translucent gland dots (becoming yellowish brown in dried specimens); margin serrate-dentate with 6–10 pairs of teeth; petiole reddish purple, 10–20(–30 mm). *Inflorescence* a short terminal raceme (30–) 70–90 (–120) mm long, sometimes with a pair of side branches at base; rachis sparsely strigose, bearing scattered, sessile, yellowish brown gland dots. *Flowers* in sessile, 1–3-flowered cymes forming 2–6-flowered verticillasters, the latter 6–12(–18) mm apart. *Calyx* up to 4 mm long, accrescent, lengthening to 10–11 mm in fruit, 2-lipped. *Corolla* pink; tube straight, 9–10 mm long, laterally compressed, 3 mm deep, slightly deflexed forming a swollen saccate base; 2-lipped; upper lip 4-lobed, 8 mm high, becoming reflexed when the stigma matures; upper lobes bent forward and forming an ascending spreading 2-spurred hood; lower lip boat-shaped, 6 mm long; soon becoming reflexed. *Style* 14 mm long extending up to 20 mm when mature, exposed for 8 mm. *Nutlets* rounded, 1.5 × 1.3 mm, dark brown, smooth.

Flowering time: Late summer into winter (March to June with a peak at the end of March).

Diagnostic features: It is distinguished by its scandent growth and short parallel-sided corolla tube 10 mm long. It appears to be closest to *P. reflexus* which is an erect shrub with a corolla tube 25 mm long and constricted at the mouth, and also with fleshy roots, whereas *P. mzimvubuensis* has distinct root tubers.

Distribution: *P. mzimvubuensis* is only known from the Mzimvubu River, in an area below Ludonga Village about 40 km from the Indian ocean, at about 600 m.

Habitat: The vegetation is arid savannah with an annual summer rainfall of 1000–1250 mm. The climate is subtropical, with hot summers, dry, sunny, frost-free winters and cool evenings. The populations of *P. mzimvubuensis* were encountered on a south-facing cliff-face in light shade, sharing their habitat with other succulents such as *Adromischus cristatus*, *Bulbine natalensis*, *Crassula cordata*, *C. cultrata*, *C. multicava* subsp. *floribunda*, *C. orbicularis*, *Cyanotis speciosus*, *Delosperma tradescantioides* and *Peperomia blanda*. Trees and shrubs in the area include *Bauhinia bowkerii*, *Celtis africana*, *Ficus burkei* and *Euphorbia tirucallii*.

History: *P. mzimvubuensis* was discovered in 2002 on a rubber canoe expedition by the author and colleagues, and is still grown at Kirstenbosch.

Cultivation: Plants are easily propagated from cuttings and grow rapidly. It is best suited to moist to dry bushveld and thicket gardens.

Plectranthus neochilus

Plectranthus neochilus Schltr. in J. Bot., Lond. 34: 394 (1896); Codd in Bothalia 11: 392 (1975); in F. S. A. 28, 4: 150 (1985). Type: Transvaal, Barberton, Rimers Creek, Galpin 968 (K, holo.; GRA; NH).

Name: Latin, Greek compound, '*neo*', new and '*chilus*', lipped and referring to its corolla.
Common names: smelly spurflower, rotstuinsalie
Synonyms: *Coleus carnosus*; *C. neochilus*; *C. Schinzii*

Description: Decumbent, succulent herb, to 200–500 mm. *Leaves* 20–50 × 15–35 mm, obovate, pubescent, glands pink, often viscid; margin crenate (5–6 pairs of teeth); base cuneate; apex obtuse. *Inflorescence* racemose, 60–150 mm long; bracts large, imbricate, deciduous, with a recurved apiculus; verticils 6-flowered. *Corolla* 12–20 mm long, light blue to blue-purple, tube deflexed, expanding to mouth. *Calyx* 6 mm, enlarging to 7 mm after flowering. *Nutlets* dark brown, 1 mm in diameter.

Flowering time: Autumn but occasionally at other times of year.

Diagnostic features: It is related to *P. caninus*; see p. 77 for the differences between them. It is also related to *P. ornatus* but the latter has a shorter inflorescence (30–90 mm).

Distribution: Widespread; recorded from Zambia, Zimbabwe, Namibia, Botswana, Swaziland, Gauteng, Limpopo Province, KwaZulu-Natal and the Eastern Cape.

Habitat: Grassveld, either in full sun or light shade, usually among rocks in well-drained sites. Annual summer rainfall is 500–1000 mm, and frost is light or absent.

History: This species was introduced to Kirstenbosch by Mr Stevenson from Plumtree, Zimbabwe (1033/31).

Cultivation: *P. neochilus* makes an attractive succulent groundcover and is probably more widely planted than any other of this group. It is extremely adaptable and flowers best in full sun. Although especially good for dry to moist bushveld gardens, it can be grown almost anywhere where frost is not severe. In the winter-rainfall Western Cape some forms are prone to fungal attack. Plants are easily grown from cuttings, which can be rooted *in situ*. The soil should preferably be composted and well-drained.

'Pietersburg'. This commonly grown form is small, succulent and compact. According to Dr. L. E. Codd it was distributed from gardens around the Voortrekker Monument at Pretoria, but its wild origin has been established as Pietersburg. It is widely used throughout South Africa as a bedding plant and flowers profusely in spring and summer, making a useful drought-resistant groundcover. It is in the Kirstenbosch collection (941/90).

'Siteki'. This plant is low-growing with light blue-purple flowers. The leaves are up to 12 mm long and are densely villous. It reaches a height of about 100 mm (1076/77 Kirstenbosch).

Plectranthus neochilus is a most useful plant. It can cover the ground in full sun, will thrive in containers, or prevent soil erosion on steep slopes. Its succulent nature makes it the ideal 'waterwise' subject

Plate 35

Plectranthus neochilus

Origin: Kirstenbosch (ex hort.) (941/90)

Plate 36

Plectranthus oertendahlii 'Uvongo'

Origin: Uvongo, KwaZulu-Natal (120/82)

Plectranthus oertendahlii

Plectranthus oertendahlii T. H. Fries jun. in Acta Hort. Gotoburg. 1: 253 (1924); Codd in Bothalia 11, 4: 411 (1975); in F. S. A. 28, 4: 160 (1985). Type: Natal – Cult. Uppsala (UPS, holo.).

Name: After I.A. Oertendahl, head horticulturist at the Uppsala University Botanical Gardens, Sweden.
Common names: silverleaf spurflower, Swedish ivy, Oertendahl's spurflower, Novemberjust, November lights, bontblaar spoorsalie

Description: Erect to decumbent perennial or annual herb, 50–200 mm high. *Branches* tomentose. *Leaves* 30–45 × 25–45 mm, broadly ovate to suborbicular, semi-succulent, sparingly tomentose with colourless glands, upper surface with silver veins, lower surface red-purple; base cuneate; petiole 10–40 mm; apex rounded to acute; margin shallowly crenate with 5–7 pairs of teeth. *Inflorescence* a terminal raceme 70–200 mm long, often with a pair of side branches, verticillasters up to 6-flowered. *Calyx* 3 mm enlarging to 6 mm after flowering. *Corolla* 8–15 mm long, white to rose-white, saccate at the base and constricted at the throat. *Nutlets* 1.2 mm, light to dark brown.

Flowering time: Autumn into winter (February to June with a peak in April).

Diagnostic features: *Plectranthus oertendahlii* is the only species with naturally variegated leaves (variegated cultivars of *P. madagascariensis* and *P. barbatus* differ in having green lower surfaces). It appears to be most closely allied to *P. ciliatus*.

Distribution: Restricted to coastal forested river valleys in southern KwaZulu-Natal from Oribi Gorge northwards to Uvongo.

Habitat: Grows on slopes of river valleys in dappled shade often among other succulent plants. Soils are well drained and humus-rich. Rain falls mainly in summer and averages between 800 and 1250 mm per annum.

History: *P. oertendahlii* was described in 1924 by Professor Fries from cultivated plants which had been in Sweden for more than 20 years. Their wild origin was not known, however, until in 1936 plants were found growing in the Oribi Gorge by Lillian Britten. Dr Codd, who identified them, thought that they may have got to Europe through the Hermannsburg Mission Society, which was established in southern Natal in 1897 and had strong links with Scandinavia. A Swedish surveyor who had been working on the development of Uvongo Township (in which he designated areas to be left as park) may have also taken plants back with him. The species was not introduced to cultivation in South Africa, however, until the author collected it for Kirstenbosch in 1977 (807/77) (see p. 21).

Cultivation: Grown mainly for its ornamental leaves, *P. oertendahlii* is an attractive pot plant much cultivated in Europe and especially Scandinavia where it is locally known as Swedish Ivy or Novemberjust. In a container it should be re-propagated annually. It makes a pretty perennial or groundcover (if densely planted) where frost is not too severe, so is best suited for moist subtropical and bushveld gardens. In cooler areas it can behave as an annual, reseeding each spring. Cuttings can be planted *in situ* during the growing season about 200 mm apart. They root fast and with ample compost and fertilizer will soon form a closed cover. Prune after flowering, before spring.

'Mzimkulu'. The form commonly grown in Europe. It is generally smaller and less attractive than 'Uvongo' (807/77 Kirstenbosch).

'Uvongo'. This form has large, broadly ovate leaves with very attractive silver-grey variegation (120/82 Kirstenbosch). It was introduced from Uvongo in southern KwaZulu-Natal.

Plectranthus oertendahlii **'Uvongo' has large silvery green leaves. It is suitable for hanging baskets or containers**

Plectranthus oribiensis

Plectranthus oribiensis Codd in Flower. Pl. Afr. 46, t. 1809 (1980); in F. S. A. 28, 4: 164 (1985). Type: Natal, Oribi Gorge, Van Jaarsv. 2198 (PRE, holo.).

Name: '*oribiensis*', from the Oribi Gorge in KwaZulu-Natal.

Common names: Oribi spurflower, Oribi spoorsalie

Description: Erect shrub to 1.5–2.5 m, tuberous roots horizontal. *Branches* hollow (a unique feature in *Plectranthus*). *Leaves* 65–190 × 40–175 mm, broadly ovate to orbicular, densely tomentose, glands colourless; base cordate; apex acute; margin dentate with 10–18 pairs of shallow teeth. *Inflorescence* simple, 200–440 mm long, often with a pair of basal branches. *Verticillasters* 5–10 flowered. Calyx 5 mm, enlarging to 8 mm after flowering. *Corolla* 7–8 mm long, purple, pilose, spurred at the base, narrowing to the throat. *Nutlets* 2 mm, dark brown.

Flowering time: Autumn (March to May with a peak during April).

***Plectranthus oribiensis* has hairy flowers and tuberous roots. It is endemic to the Oribi Gorge in southern KwaZulu-Natal**

Diagnostic features: Its hollow stems, hairy purple flowers and horizontal tuberous roots make *P. oribiensis* easily recognizable. It is related to *P. fruticosus*, which is immediately distinguished by its glabrous eglandular corolla and fibrous roots, and to *P. grallatus* which has annual aerial parts, fibrous roots and usually white flowers.

Distribution: It is only known from its type location in the Oribi Gorge in southern Kwazulu-Natal.

Habitat: On shady slopes among sandstone rocks on the margin of the subtropical forest. Soil is well drained and humus-rich. Rain falls mainly in summer and averages between 1000 and 1750 mm per annum. Frost is absent.

History: It was first collected by H. B. Nicholson in 1971, but was only described after it was collected for Kirstenbosch by the author in 1977 (806/77).

Cultivation: *P. oribiensis* is a shrub that thrives in light shade and is best suited for moist subtropical gardens. Its tuberous roots make it drought-tolerant. It is easily propagated from seed or cuttings.

Plate 37

Plectranthus oribiensis

Origin: Oribi Gorge,
KwaZulu-Natal (806/77)

Plectranthus ornatus

Plectranthus ornatus Codd in Bothalia 11: 393 (1975); in F. S. A. 28, 4: 151 (1985). Type: Ethiopia, Schimper 2. 1328 (P).

Name: '*ornatus*' = ornamental, showy.
Common names: ornamental spurflower, tuin spoorsalie, skutblaarsalie
Synonym: *Coleus comosus*

Description: Decumbent unpleasantly aromatic succulent herb, to 130–300 mm, freely branched, neat and cushion-shaped. *Branches* somewhat 4-angled and covered with glandular hairs. *Leaves* succulent with petiole to 10 mm long, the blade 20–30 × 15–25 mm, broadly obovate, pubescent; margin crenate-dentate (4–6 pairs of teeth); base cuneate; apex obtuse. *Inflorescence* terminal, an erect dense spike-like raceme 40–90 mm long; bracts large, ovate-acuminate to 12 × 8 mm, imbricate, deciduous. *Flowers* in dense 6-flowered verticils. *Corolla* 20–25 mm long, light blue to blue-purple, tube slightly deflexed, expanding to the mouth; upper lip 6 mm long, 2-lobed and bearing two small lateral lobes; lower lip boat-shaped; stamens united at the base. *Calyx* 3 mm, enlarging to 6 mm after flowering. *Nutlets* dark brown, 2 mm long.

Plectranthus ornatus is closely related to _P. neochilus_. It is a handy drought-tolerant groundcover which can be grown in full sun

Flowering time: Throughout summer with a peak in autumn.

Diagnostic features: It is related to *P. neochilus*. Distinguishing characters are discussed under that species.

Distribution: *P. ornatus* is widespread in Central Africa from Ethiopia in the north to Tanzania in the south. It is semi-naturalized in South Africa.

Habitat: It occurs in grassland above the forest zone.

History: The earlier epithet '*comosus*' could not be transferred from the genus *Coleus* because it existed as a synonym for *P. barbatus*. *P. ornatus* has been cultivated at Kirstenbosch for a number of years under the name *P. neochilus*.

Cultivation: *P. ornatus* is a popular groundcover in gardens. It will grow in most parts of South Africa where frost is not severe and is useful in rockeries. Plants are easily raised from cuttings, which should preferably be planted in the spring.

Plate 38

Plectranthus ornatus

Origin: Kirstenbosch (ex hort.)

Plectranthus pentheri

Plectranthus pentheri (Gürke) Van Jaarsv. & T. J. Edwards in Bothalia 27 (1): 5–6, ill. (1997). Type: RSA, Eastern Cape Krook s.n. in Penther 1716 (W. PRE)

Name: After Arnold Penther (1865–1931), zoologist and naturalist who collected in South Africa during 1894–5.

Common names: Penther spurflower, Penther spoorsalie

Synonym: *Coleus pentheri*

Description: Decumbent succulent herb to 100 mm; basal tuber 30 × 15 mm, white. *Stems* 4-angled, 2–4 mm in diameter, sparsely pilose, punctate, glands orange; internodes 3–6 mm long. *Leaves* conduplicate, succulent, firm, ovate to subrotund, 9–14 × 9–14 mm, entire to dentate-serrate in upper half, teeth shallow, 4 or 5 pairs; sparsely pubescent, punctate, glands orange, base broadly cuneate, apex obtuse; petiole 2 mm long. *Raceme* 50–70 mm long, simple; cymes 3-flowered, 5–10 mm apart; bracts broadly ovate to subrotund, mucronate, 4 × 4 mm, initially imbricate, caducous; pedicels 5 mm long. *Calyx* 3 mm long (enlarging to 5 mm), upper lip broadly ovate, 2 × 2.5 mm, lower lobes 4, linear, 2 mm long. *Corolla* 15 mm long, white; tube laterally compressed, slightly geniculate, 5 × 1.5 mm, expanding to 3 mm at the throat, upper lip 3 mm long, 2-lobed, lateral lobes 2.5 mm long, lower lip boat-shaped, 9 mm long. *Nutlets* brown to black, ovoid, 1.5 × 1 mm.

Flowering time: Autumn (March to May).

Diagnostic features: *P. pentheri* is closely related to *P. neochilus*, but has oblong tuberous roots and white flowers, and also lacks the strong aroma of the widespread *P. neochilus*, which has fibrous roots and mauve to purple flowers.

Distribution: Eastern Cape to KwaZulu-Natal.

Habitat: Plants grow on shallow soil among granite boulders in grassland.

History: *P. pentheri* was first collected by Arnold Penther at Breakfast Vlei in the Eastern Cape and described by the botanist Gürke in 1905.

Cultivation: Easily grown from cuttings. Best suited to rockeries in sunny areas of subtropical or bushveld gardens.

Plate 39

Plectranthus pentheri

Origin: Eastern Cape, near the Kei River (458/95)

Plectranthus petiolaris

Plectranthus petiolaris E. Mey. ex Benth. in E. Mey., Comm. 228 (1837); in DC., Prodr. 12: 66 (1848); Codd in Bothalia 11: 431 (1975); in F. S. A. 28, 4: 171 (1985). Type: Transkei, between Umtata and Mzimvubu Rivers, Drège 4773b (K.).

Name: '*petiolaris*' = with a leaf-stalk, referring to the long leaf-stalks.
Common names: Dutch-pipe spurflower, kromsteelpyp spoorsalie
Synonym: *Plectranthus kuntzei*

Description: Decumbent to scandent herb to 300–600 mm, stems brittle, succulent. *Leaves* membranous, 20–100 × 20–100 mm, blade broadly ovate, surface sparingly strigose, glands colourless; base cordate or truncate, petiole 30–120 mm long; apex acute or obtuse; margin dentate with 8–10 or more pairs of teeth. *Inflorescence* a terminal raceme up to 250 mm long, often with a pair of basal branches; verticillasters 4 to 6-flowered. *Corolla* 10–17 mm long, violet-purple to pink, deflexed and widening to the throat. *Calyx* 2.5 mm, enlarging to 7 mm after flowering. *Nutlets* ±1.75 mm long, brown.

Flowering time: Autumn (March to April).

Diagnostic features: The sigmoid 'Dutchman pipe' corolla shape is characteristic of a number of African plectranthus, including two other South Africans. *P. laxiflorus* is distinguished by its white flowers with honey-coloured markings, and *P. dolomiticus* has smaller leaves and tuberous roots.

Distribution: Widespread from the northern part of the Eastern Cape to Mpumalanga. In KwaZulu-Natal it is mainly associated with coastal regions although it does occur in Crocodile Gorge near Malelaan.

Habitat: Semi-shaded areas of subtropical coastal forest (and bushveld) in low-altitude river valleys. It is often found on screes and frequently occurs as a lithophyte in forests. Soils are humus-rich and well drained. Rain falls mainly in summer and ranges between 800 and 1500 mm a year.

History: It was first collected in 1832 between the Umtata and Mzimvubu Rivers by Johann Drège. It was introduced to Kirstenbosch in 1977 (771/77).

Cultivation: *P. petiolaris* is best suited for shade in warm subtropical or frost-free bushveld gardens, and also does well at the seafront. It is easily grown from cuttings (which can be planted 400 mm apart), developing into a vigorous groundcover that layers freely and can be rampant. It tolerates poor soils. Can be pruned.

Plate 40
Plectranthus petiolaris
Origin: Crocodile Gorge, Mpumalanga (771/77)

Plectranthus porcatus

Plectranthus porcatus Winter & Van Jaarsv. in Bothalia 35, 2: 169–173. Type: Sekhukhuneland, northern Leolo Mountain range, Farm De Kamp. P. J. D. Winter 6725 (PRE, holo.).

Name: The epithet '*porcatus*' alludes to the ridged nature of most stem internodes. These ridges are absent on some internodes during stages of active growth.
Common names: Ribbed spurflower, Sekukuni spoorsalie

Description: Perennial soboliferous multi-stemmed aromatic shrub to ± 1.2 m and 1.5 m in diameter, aerial stems sparingly branched. *Roots* succulent, to 14 mm across. *Stems* herbaceous, semi-succulent, prominently 4-angled, laterally compressed decussately below the nodes (distinctly rectangular, 10 × 8 mm in transverse section), with a pair of longitudinal ridges from each leaf base, decurrent along angles on each side; surface hairy with sessile orange glandular trichomes; bark of older branches rusty brown becoming scabrid. *Leaves* broadly ovate to ovate-deltoid, 55–120 × 40–90 mm; apex obtuse to acute; base shortly cuneate and decurrent on petiole; margin serrate-dentate to serrate-crenate with 10–12 pairs of teeth; surface glandular pubescent; petiole ascending, 7–25 mm long. *Inflorescence* a short terminal raceme, 200–230 mm long, and with a pair of side branches at base carried on short peduncles to 50 mm long; rachis reddish-brown, bracts triangular-ovate 5 × 4 mm, early deciduous. *Flowers* in 7-flowered fascicles forming 14-flowered verticillasters. *Corolla* 2-lipped, somewhat sigmoid, 16 mm long, white to slightly mauve-tinged, glandular hairy and sparsely dotted with sessile orange glandular trichomes; upper lip erect, unequally 4-lobed, the two median lobes the larger and 2 mm high. *Stamens* 4, didynamous, filaments shortly connate at the base; anthers versatile, 0.8 mm long, purple, becoming black; pollen yellow. *Calyx* 4–6 mm long, lengthening to 8 mm at maturity. *Nutlets* 1.8 mm, brown, smooth.

Flowering time: Autumn (March to April).

Diagnostic features: *P. porcatus* is at once distinguished by its unique, peculiarly ribbed stems which are rusty brown and scabrid in mature bark, as well as by its stoloniferous nature. It is related both to *P. xerophilus* (with purple flowers) and *P. unguentarius* (with greenish white flowers), both of which have smooth bark.

Distribution: Northern Leolo Mountains, Sekukuniland, Limpopo Province.

Habitat: Dry *Lydenburgia cassinoides* and *Kirkia wilmsii* bushveld, in light shade of shrubs or among boulders, usually on the south-west side. Rainfall is 400–600 mm per annum and occurs mainly from summer to autumn. The climate is subtropical, with hot summers and dry, sunny winters with light frost. The plants grow in dark brown loam derived from gabbro-norite rocks of the Bushveld igneous complex.

History: *P. porcatus* was discovered by Pieter Winter of the National Herbarium in Pretoria whilst searching for more populations of a new species of *Raphionacme* in the Leolo Mountains.

Cultivation: Plants can be grown in full sun or light shade, and are best suited to bushveld or even highveld gardens with a little protection. They multiply by means of their subterranean stolons and are easily grown from cuttings. They do well in containers.

Plate 41

Plectranthus porcatus

Origin: Collected by Pieter Winter on the Leolo Mountains, Sekukuniland, Northern Province (495/06)

Plectranthus praetermissus

Plectranthus praetermissus Codd Flower. Pl. Afr. 45: t. 1791 (1979); in F. S. A. 28, 4: 160 (1985). Type: Transkei, Port St Johns. Stutterheim sub PRE 57330 (PRE, holo.).

Name: Latin, '*praetermissus*' = overlooked, referring to its late discovery.
Common names: Port St Johns spurflower, Port St Johns spoorsalie

Description: Procumbent or decumbent herb (occasionally annual), 100–300 mm high. *Stems* purplish, sparingly pubescent. *Leaves* 25–90 × 20–85 mm; ovate to broadly ovate, sparingly pubescent, glands red; base rounded; apex acute or obtuse; margin dentate with 5–6 pairs of teeth, often undulate. *Inflorescence* terminal, racemose, lax, 80–250 mm long. *Calyx* 5 mm long, elongating to 10 mm after flowering. *Corolla* 15–17 mm long, purple-violet; tube straight, saccate at the base and gradually narrowing to the throat. *Nutlets* 1.5 mm, black or dark brown.

Flowering time: Autumn (February to May with a peak in late March and early April).

Diagnostic features: In his original description Codd likened *P. praetermissus* to *P. oertendahlii*, but *P. praetermissus* can easily be distinguished by its mauve-purple flowers and non-variegated leaves. Prostrate forms are vegetatively similar to *P. verticillatus*, and these species share similar red glands, but plants are easily distinguished by flower colour, the latter being white. *P. praetermissus* is allied to *P. reflexus*, however, which is an erect shrub with semi-succulent leaves, dark purple on the undersurface, and a narrow corolla ±27 mm long. *P. ambiguus* is superficially similar but has a long narrow, parallel-sided corolla tube 20–30 mm long and 1 mm wide.

Distribution: Known only from near Port St Johns in the Eastern Cape.

Habitat: Moist semi-coastal subtropical forest, often in association with *Impatiens flanaganiae*, *Begonia sutherlandii* and a few other *Plectranthus* species. Rainfall occurs throughout the year and the annual average is 1500–1750 mm.

History: *P. praetermissus* was first collected in 1921 by the plant pathologist and frog collector Dr Vincent Wager. It was again collected in 1925 by Mrs Howlett, but it was only after Mrs Nico Stutterheim collected live material in 1979 that its status became clear. It was introduced to Kirstenbosch in 1979 (63/79).

Cultivation: Its striking mauve flowers are stunning. It makes an attractive groundcover for shady subtropical gardens, and also thrives as a pot plant or in a hanging basket. It is best to re-propagate annually, as plants tend to become moribund after a few seasons.

***Plectranthus praetermissus* does well in hanging baskets. It flowers during autumn**

Plate 42

Plectranthus praetermissus

Origin: Port St Johns, Transkei, Eastern Cape (63/79)

Plate 43

Plectranthus psammophilus

Origin: Makatini Flats, KwaZulu-Natal (767/77)

Plectranthus psammophilus

Plectranthus psammophilus Codd in Bothalia 11: 405 (1975); in F. S. A. 28, 4: 156 (1985). Type: Natal, Makatini Flats, Strey 5779 (PRE, holo.).

Name: Greek '*psammos*', sand and '*phileo*', to love, referring to its sandy habitat.
Common names: Makatini spurflower, Makatini spoorsalie.

Description: Prostrate or decumbent semi-succulent herb, bearing branches up to 700 mm long. *Branches* 4-angled with glandular hairs and covered with small red glands. *Leaves* with petioles up to 20 mm long; the blade 15–40 × 14–40 mm, ovate-triangular, semi-succulent, the surface sparingly pubescent and covered with minute reddish-brown glands; base truncate; apex acute; margin crenate-dentate with 5–7 pairs of teeth. *Inflorescence* terminal, racemose 100–300 mm long, occasionally with a pair of side branches; rachis with glandular hairs. *Flowers* in 3–6-flowered cymes forming 6–12-flowered verticillasters. *Corolla* ±5 mm long, purple-blue; tube centrally deflexed, widening to the throat; upper lip erect, 2-lobed bearing two smaller lateral lobes; lower lip boat-shaped, horizontal, 2.5 mm long. *Calyx* 1.5 mm enlarging to 3 mm after flowering. *Nutlets* 1 mm, light to dark-brown.

Flowering time: Autumn (March to May with a peak in April).

Diagnostic features: *P. psammophilus* is related to *P. madagascariensis* which has a larger white corolla 7–18 mm long.

Distribution: Restricted to the sandy Makatini Flats in northern KwaZulu-Natal.

Habitat: Savannah vegetation in light shade; annual rainfall is 600–1250 mm and frost is rare. Where it occurs it is locally abundant. It is found in association with *Aloe greenii*, *Cissus rotundifolia*, *Kalanchoe rotundifolia*, *Sarcostemma viminale*, *Senecio barbertonicus*, *S. viminalis* and *Sanseviera hyacinthoides*.

History: *P. psammophilus* was introduced to Kirstenbosch in 1977 (767/77).

Cultivation: Easily propagated from cuttings or seed but the plants are rather insignificant. It is suitable for dry sandy bushveld gardens. It can be prone to downy mildew.

Plectranthus purpuratus

Plectranthus purpuratus Harv. in Thes. Cap. 1: 53, t.83 (1859); Codd in Bothalia 11, 4: 410 (1975); in F. S. A. 28, 4: 158 (1985). Type: Natal, ex hort. Kew, from seed sent from Port Natal by R. Vause s.n. (K).

Name: '*purpuratus*', made purple, referring to the purple underside of the leaf; '*montanus*', from a mountain habitat; '*tongaensis*', from the Tongaland region of KwaZulu-Natal.

Common names: purple-leaved spurflower, cliff spurflower, persblaar spoorsalie, krans spoorsalie

Description: Erect to decumbent spreading herb; stems 30–100 mm long. *Leaves* 15–35 × 15–35 mm, ovate to broadly ovate, succulent, often purplish below, densely to sparingly strigose; glands red; base truncate to cuneate; apex rounded or acute; margin shallowly crenate. *Inflorescence* terminal, racemose, 30–120 mm long, often with a pair of side branches; verticillasters up to 8-flowered. *Calyx* 3 mm enlarging to 5–6 mm after flowering. *Corolla* 10–11 mm long, white to bluish, tube saccate at the base and constricted near the middle. *Nutlets* 1 mm, light to dark brown.

Flowering time: September–May, with a peak in April.

Diagnostic features: Closely related to *P. strigosus* but distinguished by its corolla tube which is constricted in the centre. The leaves of *P. strigosus* are usually rusty, strigose and the corolla tube is constricted at the throat. Codd (1975, 1985) considers the northern forms of *P. purpuratus* (from northern KwaZulu-Natal, Swaziland and Mpumalanga) as part of the *P. strigosus* complex. Although *P. purpuratus* is variable, the distinctive medial constriction of the corolla tube, in my opinion, justifies its separation from *P. strigosus*.

P. purpuratus has been divided into three subspecies.

Subsp. *montanus* has a procumbent habit. Its leaves are ovate to broadly ovate, 12–15 × 10–15 mm, green to purplish with 2 pairs of shallow teeth. It is covered with fine grey hairs.

Subsp. *purpuratus* has compact growth with erect to decumbent stems. Its leaves are subimbricate, subrotund to broadly ovate, succulent, 15–45 × 15–38 mm, grey-green, entire, occasionally shallowly crenate, with 3 pairs of teeth, subglabrous, rubropunctate beneath; apex rounded; base truncate to cuneate.

Subsp. *tongaensis* is also procumbent, rooting at the nodes. Its leaves are ovate to broadly ovate, 30–40 × 15–18 mm, green to purple-tinged, occasionally serrate with 3–4 pairs of shallow teeth.

Flowering time: Autumn (March to May).

Distribution: Subsp. *purpuratus* is confined to the Durban–Pietermaritzburg region of KwaZulu-Natal. Subsp. *montanus* is from the southern Mpumalanga and Swazi Drakensberg, where it occurs on the mountain peaks (1300–1700 m). Subsp. *tongaensis* is restricted to Tongaland, from St. Lucia to Kosi Bay.

Habitat: *P. purpuratus* subsp. *purpuratus* grows on rocky outcrops and south-facing cliffs in bushveld. Subsp. *montanus* occurs on rocky mountains in grassland, and subsp. *tongaensis* in sandy coastal forest margins.

History: Seed of subsp. *purpuratus* was sent to Kew by Mr R. Vause; the plant was named by William Harvey, Colonial Treasurer at the Cape, in 1859. It was introduced to Kirstenbosch in 1977 (775/77). Subsp. *montanus* was first collected by Compton near Mbabane in Swaziland, and was introduced to Kirstenbosch in 1982 (114/82). Subsp. *tongaensis* was collected in 1992 near Kosi Bay (275/93).

Cultivation: Easily propagated from cuttings. Subsp. *purpuratus* does well in containers and is also suited for rockeries in bushveld gardens. Subsp. *montanus* and subsp. *tongaensis* both make useful groundcovers, though the latter is better in coastal subtropical gardens.

***P. purpuratus* subsp. *montanus* 'Lomati'.** A densely hairy cultivar from Lomati Gorge near Barberton, Mpumalanga (1074/77). This alpine is resistant to light frost.

***P. purpuratus* subsp. *purpuratus* 'Umgeni'.** This cultivar has larger leaves suffused with dark purple on the undersurface. It originates from the Umgeni Valley at Durban (898/88 Kirstenbosch).

***P. purpuratus* subsp. *tongaensis* 'Kosi'.** A mat-former with dark purple undersides. Suitable as a groundcover especially in coastal and seafront gardens (275/93). This collection originates from Kosi Bay in northern KwaZulu-Natal.

Plate 44

Plectranthus purpuratus
subsp. *purpuratus*

Origin: Umgeni Valley, Durban, KwaZulu-Natal (898/88)

The left stem illustrates a plant grown in sun; the central stem a plant grown in shade

(*Plectranthus purpuratus* **cont.**)

Plate 45

Plectranthus purpuratus

A: ***P. purpuratus*** subsp. ***tongaensis***

Origin: Kosi Bay, KwaZulu-Natal

B: ***P. purpuratus*** subsp. ***montanus***

Origin: Barberton, Mpumalanga (898/88)

Plectranthus ramosior

Plectranthus madagascar var. *ramosior* Benth. In DC., Prodr. 12: 68 (1848). Type: Magaliesberg, Burke (K; BM).

Name: *ramosior* = more branched
Common names: Magaliesberg spurflower, Magaliesberg spoorsalie
Synonyms: *P. madagascariensis* var. *ramosior*

Description: Perennial semi-succulent herb up to 120 mm high, rootstock tuberous, branches decumbent. *Leaves* 35–55 × 20–35 mm, ovate to broadly ovate, puberulous to tomentose; base truncate to cuneate; apex rounded to subacute; margin crenate-dentate with 4–7 pairs of teeth. *Inflorescence* terminal, racemose up to 120 mm long, occasionally with a pair of side branches; flowers in 4–8-flowered verticillasters, 10–30 mm apart. *Corolla* 8–12 mm long, mauve (rarely white); tube deflexed, widening towards the throat. *Calyx* 2–3 mm, enlarging to 5–6 mm after flowering. *Nutlets* 1.2 mm, brown.

Flowering time: Autumn into winter (March to June with a peak in April).

Diagnostic features: *Plectranthus ramosior* is closely related to *P. hadiensis*, *P. madagascariensis, P. grandidentatus* and *P. woodii*. From *P. hadiensis*, *P. madagascariensis* and *P. grandidentatus* it is at once distinguished by its tuberous roots. *P. woodii* also has tuberous roots but its leaves are much larger and its flowers are usually white, while the plants are confined to the Drakensberg. *P. ramosior* was formerly considered a variety of *P. madagascariensis*, but on the grounds of its mauve or bluish corolla and tuberous roots it is now considered a separate entity.

Distribution: Witwatersrand and Magaliesberg (Gauteng) and further west and north to the Waterberg (Limpopo Province).

Habitat: It is a species of dry savannah almost always found on rocky outcrops and cliffs. The plants are subject to seasonal frost and veld fires and re-sprout annually from their persistent tuberous rootstocks. The soil is well drained. Rainfall ranges from 600 to 1000 mm per annum.

History: *P. ramosior* was first collected by Joseph Burke (1812–73), a well-known English plant collector. It was introduced to Kirstenbosch in 1982 from Rooiwal near Greylingstad (Gauteng) (121/82).

Cultivation: This tuberous-rooted species with its mauve to bluish flowers can be grown in full sun, and would suit a rockery in a grassland or sunny highveld garden. It is easily propagated from cuttings and grows rapidly; young plants can be set out 300 mm apart, and should flower the same year. It is best to prune after flowering.

Plate 46

Plectranthus ramosior

Origin: Kransberg, Northern Province

Plectranthus reflexus

Plectranthus reflexus Van Jaarsv. & Edwards in Flower. Pl. Afr. 51(2) t.2034. Type: Cape Province – 'Bulolwe River south of Port St Johns'. Van Jaarsv. Campher & Dekker 10156 (NBG, holo.!).

Name: Latin, '*reflexus*' = reflexed. The lower corolla lobe becomes distinctly reflexed with age.
Common names: ballerina spurflower, ballerina spoorsalie

Description: Soft, erect, sparingly branched herb to 400–800 mm (rarely a shrub to 1500 mm); stems purplish, becoming terete and basally swollen, strigose-pubescent, with scattered, long multicellular hairs. *Leaves* 40–70 × 35–50 mm, triangular-ovate, sparingly pubescent, glands colourless; lower surface suffused with purple; base rounded; apex acute; margin dentate. *Inflorescence* terminal, racemose, 150–200 mm long. *Calyx* 4 mm, enlarging to 8 mm after flowering. *Corolla* 27–28 mm long, purplish; tube 25 mm long, compressed, straight in the proximal half, 3 mm wide at the base and narrowing to 1.5 mm at the throat. *Nutlets* 1.5 mm long, black or dark brown.

Flowering time: Midsummer into autumn (January to March).

Diagnostic features: This recently discovered herb is allied to *P. praetermissus*, but the latter is a decumbent herb with a much shorter corolla. It can also be confused with *P. ambiguus*, *P. ecklonii* and *P. hilliardiae*, all of which have fairly long straight corolla tubes. *P. reflexus* can be distinguished by its corolla which is saccate in the proximal half and by its semi-succulent, triangular-ovate leaves which are purple on the abaxial surface. The basally swollen stem base is often supported by stilt roots, another unique feature. *P. ambiguus* has a long (20–30 mm), narrow, parallel-sided corolla tube 1 mm wide, while *P. hilliardiae* has a tube 4 mm across. *P. ecklonii* is a tall shrub with short corolla tubes (12–15 mm long) which dilate evenly to the throat.

Distribution: *P. reflexus* is known only from near Port St Johns in the Eastern Cape.

Habitat: It chiefly grows on sandstone rocks of the Cape Supergroup (Natal series) in coastal forest. Associated plants include *Stangeria eriopus*, *Smilax kraussiana*, *Flagellaria guineensis* and *Carissa wyliei*. Winters tend to be dry, though rainfall can occur throughout the year, averaging 1500–1700 mm per annum.

History: *P. reflexus* was first collected in 1989 along the Bulolwe River near Port St Johns by the author who introduced it to Kirstenbosch (460/89). For an account of the expedition, see pp. 21-2.

Cultivation: *P. reflexus* make an attractive shrub, but is not as easy as most plectranthus; it tends to be slow and rather lacking in vigour. It is propagated by cuttings, and young plants can be lightly pruned to encourage bushiness. Recommended for humid, subtropical gardens where it will tolerate the occasional dry spell.

***Plectranthus reflexus* is an erect shrub endemic to the northern part of the Eastern Cape, occurring from Port St Johns to the Msikaba River**

Plate 47

Plectranthus reflexus

Origin: Port St Johns, Eastern Cape (460/89)

Plate 48

Plectranthus rehmannii

Origin: Karkloof, near Pietermaritzburg, KwaZulu-Natal

Plectranthus rehmannii

Plectranthus rehmannii Gürke in Bull. Herb. Boissier 6: 553 (1898); Codd in Bothalia 11: 421 (1975); in F. S. A. 28, 4: 166 (1985). Type: Natal, Karkloof, Rehmann 7359 (Z holo., K).

Name: After Anton Rehmann (1840–1917), a botanist and geographer of German and Austrian descent.
Common names: Karkloof spurflower, Rehmann spurflower, Karkloof spoorsalie, Rehmann spoorsalie

Description: Erect freely branched aromatic shrub up to 1.5 m. *Branches* 4-angled bearing short hairs. *Leaves* 25–140 × 12–80 mm with petioles up to 60 mm long; blade ovate to broadly ovate, the surface sparingly pubescent, the lower side reticulately veined and covered with minute reddish glands; base truncate; apex acute or obtuse; margin crenate-serrate with 25–50 pairs of teeth; glands red. *Inflorescence* a terminal panicle, 250–350 mm with 1–3 pairs of branches near the base; rachis covered with short hairs; bracts ovate-lanceolate and persisting beyond the flowering stage. Flowers in 3–8-flowered cymes forming up to 16-flowered verticillasters, each up to 20 mm apart. *Corolla* 6–7 mm long, white, densely villous; tube deflexed, saccate at the base narrowing to the throat; upper lip ascending, broadly ovate, up to 3 mm long; lower lip 5 mm long. *Calyx* 2.5 mm long, enlarging to 9 mm after flowering. *Nutlets* 1.5 mm, black or brown.

Flowering time: Autumn into winter (March to May).

Diagnostic features: Its hairy white corolla distinguishes it from all other African plectranthus.

Distribution: Endemic to the afromontane mist-belt forests of the KwaZulu-Natal midlands.

Habitat: Forest margins and stream banks where the annual rainfall is between 800 and 1000 mm, falling predominantly in summer. In its natural habitat this species is often deciduous. Plants occur in association with *Plectranthus fruticosus, P. laxiflorus, P. ciliatus* and *P. ambiguus*. Other forest associates include *Dracaena aletriformis, Peperomia retusa, Senecio tamoides* and *Streptocarpus species.*

History: It was introduced to Kirstenbosch in 1932 by Mr Carnegie from Pietermaritzburg.

Cultivation: *P. rehmannii* is easily raised from cuttings or seed. It develops into an erect shrub suitable for shady or semi-shady areas of frost-free subtropical forest gardens. It should be pruned after flowering and before spring. Although not as spectacular as many of its relatives, its curiously fuzzy white flowers have a distinctive charm.

Plectranthus rehmannii **has distinctive hairy white flowers. It makes an erect shrub**

Plectranthus rubropunctatus

Plectranthus rubropunctatus Codd in Bothalia 11, 4: 420 (1975); in F. S. A. 28, 4: 165 (1985). Type: Transvaal, Barberton district. Nelshoogte Forest Station, Strey 4081 (PRE, holo.).

Name: Latin, '*rubro*', red and '*punctatus*', spotted, referring to the red gland dots on the underside of the leaves.
Common names: red-dot spurflower, Mpumalanga spurflower, Mpumalanga spoorsalie, robyn spoorsalie

Description: Erect shrub or procumbent herb up to 2 m high, stems purplish. *Leaves* 35–140 × 25–100, broadly ovate, chartaceous, hispidulous, with red glands; base alternate to cordate; apex acute to rounded; margin crenate with 12–30 pairs of teeth. *Inflorescence* a terminal panicle, 150–200 mm long, verticillasters usually 6-flowered. *Calyx* 6–10 mm, enlarging to 12 mm after flowering. *Corolla* 6–10 mm long, white; saccate at the base narrowing to the throat. *Nutlets* ±1.75 mm, dark brown.

Flowering time: Autumn into winter (March to June with a peak during April).

Diagnostic features: Allied to *P. grallatus* which differs in having annual aerial stems and a tuberous rootstock. Two forms of *P. rubropunctatus* are recognized: an erect shrub associated with forest shade, and a procumbent form usually associated with moist grassland. Floristically it is also related to *P. fruticosus*, which is distinguished by its spurred corolla, and to *P. oribiensis*, which has hairy purple flowers.

Distribution: Chiefly restricted to the Drakensberg escarpment of Swaziland, Mpumalanga and Limpopo Province, but also found on the Zoutpansberg and Blouberg Mountains in Limpopo Province.

Habitat: Moist forest, shrub forest or exposed forest margins, usually in the shade on well-drained and humus-rich soils derived from sandstone. Rain falls predominantly in summer and averages 1500–2000 mm a year. Frost is light or absent.

History: *P. rubropunctatus* was originally collected by the Swiss missionary, Henri-Alexandra Junod (1863–1934). It was first cultivated at the Pretoria National Botanic Gardens in 1950, and was introduced to Kirstenbosch in 1976 (776/75).

Cultivation: It grows well under moist shady conditions and is best suited to forest gardens. Procumbent forms can be used as groundcovers in partial or full shade. All forms are easily propagated from cuttings, which should be planted about 300 mm apart.

'Entabeni'. Erect floriferous shrub 1–1,5 m with white to light mauve flowers.

'Mariepskop'. Groundcover with white flowers.

Plate 49

Plectranthus rubropunctatus

Origin: Near Graskop, Mpumalanga

Plectranthus saccatus

Plectranthus saccatus Benth. in E. Mey., Comm. 227 (1837); Codd in Bothalia 11, 4: 426 (1975); in F. S. A. 28, 4: 168 (1985). Type: Transkei, Mzimvubu River, Drège 1675 (K, ex Herb. Benth., holo.; G; P; S).

Name: Latin, '*saccatus*', bag-shaped or pouched, referring to the base of the flower.
Common names: stoep jacaranda, stoep jakaranda, sak-spoorsalie

Description: Procumbent, decumbent or erect aromatic shrub up to 2 m, stems puberulous. *Leaves* 15–60 mm × 12–45 mm, blade thin to succulent, broadly ovate to triangular, sparingly strigose to glabrescent, glands colourless; base obtuse or truncate; apex acute; margin crenate to dentate with 3–6 pairs of teeth. *Inflorescence* a terminal raceme 25–120 mm long, verticillasters 2–6 flowered. *Calyx* 3–4 mm, enlarging to 8 mm after flowering. *Corolla* 11–30 mm long, mauve to blue (rarely white); tube saccate at the base, compressed, parallel-sided, 5–8 mm broad. *Nutlets* 2 mm, dark brown.

Flowering time: Early summer to late autumn (November to June with a peak during April, also sporadically throughout the year).

Diagnostic features: Two subspecies are recognized. **Subsp. *saccatus*** is a decumbent to erect shrub with a widespread distribution and a large saccate corolla. Its leaves are also variable but usually have a thin and sometimes slightly succulent blade. Florally it appears allied to *P. hilliardiae*, but the large corolla with its upper lip of 10 to 16 mm long is a unique feature amongst South African species. Its variety *longitubus* has a long corolla tube of 20–26 mm, and var. *saccatus* a tube of about 14–20 mm long.
P. saccatus subsp. *saccatus* is very variable with several local forms or ecotypes. The recognition of two varieties within this subspecies is contentious, and a careful consideration of corolla lengths shows that they are highly variable, representing a continuum rather than forming clearcut groups within the species.

Subsp. *pondoensis* (Pondo stoep jacaranda) is distinguished by its trailing habit, and its flexible stems up to 4 m long bearing distinctly succulent glandular pubescent leaves 12–27 × 10–28 mm. It usually has a short corolla tube 6–10 mm long, compared to the much larger corolla of subsp. *saccatus*. It tends to flower throughout the year with a peak in autumn but is never as floriferous as subsp. *saccatus*.

Distribution: From the Eastern Cape to northern KwaZulu-Natal. Subsp. *pondoensis* is endemic to quartzitic sandstone gorges from Pondoland north to Oribi Gorge.

Habitat: Forest or forest margins, in hilly terrain or deep river gorges. Subsp. *pondoensis* is confined to steep slopes and sheer cliff-faces where the plants occasionally droop over the rockface. Soils are humus-rich and fertile. Rainfall averages 1000–1500 mm per annum, falling mainly in summer, and frost is absent.

History: *P. saccatus* was first collected in 1832 by Drège, a German botanical collector and traveller. Medley Wood, who collected it in Ngoye Forest, rated it highly. In his *Natal Plants* (1899) he wrote that of the many Natal plectranthus, some of which were 'as yet undecided', *P. saccatus* was 'one of the most attractive', adding that it was 'usually found in damp places in shade and frequently seen in cultivation'. He sent plants to the Cambridge University Botanic Garden, and one that flowered there was illustrated in *Curtis's Botanical Magazine* in 1902. The species was introduced to Kirstenbosch in 1915 (4195/15). Subsp. *pondoensis* was first collected by Mr Marais from the Mkambati River in the Eastern Cape, and received by Kirstenbosch in 1977 (808/77).

Cultivation: With its large mauve to white flowers (the largest in the genus), *P. saccatus* is one of the showiest and most sought-after species, and is a popular shrub in many countries. It is best suited for subtropical forest gardens, and will make a good shrubby groundcover. In humid areas it will tolerate nearly full sun. Procumbent forms can be used to cover the ground in shade or partial shade. It also does well as a pot plant; in Pretoria, according to Dr Codd, it is known as 'stoep jacaranda'. Subsp. *pondoensis* will tolerate full sun in regions close to the coast. This is a handy long-lived perennial and excellent groundcover, useful on steep south-facing embankments and gabions.

***Plectranthus saccatus* subsp. *pondoensis*:**
'Myokane White'. A form with white flowers spotted with mauve which will tolerate full sun in areas close to the coast.

Plate 50

Plectranthus saccatus

A: ***P. saccatus*** subsp. ***saccatus*** 'Kirsten'. Origin: Unknown (Kirstenbosch 4195/15)

B: ***P. saccatus*** subsp. ***saccatus*** 'Nkandla'. Origin: Nkandla Forest, KwaZulu-Natal (1303/82)

C: ***P. saccatus*** subsp. ***saccatus*** 'Izingolweni'. Origin: Oribi Gorge, KwaZulu-Natal

D: ***P. saccatus*** subsp. ***pondoensis***. Origin: Oribi Gorge, KwaZulu-Natal (EvJ2201; 808/77)

(*Plectranthus saccatus* cont.)

P. saccatus **subsp.** ***saccatus*****:**

'Gwaloweni'. A soft erect shrub to 1 m with large thin leaves (40–50 mm long) and a long pale mauve corolla of 20–26 mm. This form is from the Gwaloweni Forest on the Lebombo mountains in northern KwaZulu-Natal. It will take deep shade (764/77 Kirstenbosch).

'Izingolweni'. A prostrate form with heavy purple speckling in the flower, introduced from Oribi Gorge in southern KwaZulu-Natal. This form thrives in hanging baskets where it will need annual replanting (800/77 Kirstenbosch).

'King Goodwill'. Named in honour of the Zulu King Goodwill, this selection originates from deep within Ngoye Forest. It has a striking white corolla with a tube 25 mm long and shiny dark green leaves. It thrives in gardens, making a shrub of 1.5 m (344/93 Kirstenbosch).

'Kirsten'. A shrub of unknown provenance, to 1.8 m, with dark purplish-blue flowers (4195/15 Kirstenbosch).

'Mtamvuna'. A semi-succulent floriferous decumbent shrub to 300–500 mm. The corolla has a long tube and is heavily speckled with purple. It originates from the Mtamvuna River on the southern border of KwaZulu-Natal, and was introduced to Kirstenbosch in 1979 (116/82 Kirstenbosch).

'Mtunzini'. This white-flowered shrub was donated to Kirstenbosch in 1980 by Ian Garland from Twin Streams in Mtunzini. It often appears at the annual Kirstenbosch plant sale and is much grown in Cape Town gardens (423/82 Kirstenbosch). To 1 m.

'Ngoye'. Similar to 'King Goodwill' but with mauve flowers. It also originates from Ngoye Forest (345/93 Kirstenbosch).

'Nkandla'. An erect shrub to 1 m with the upper lip of the corolla not reflexed but forming a hood. The flowers are mauve, lined with darker stripes. It is suitable for frost-free shady gardens or as an indoor pot plant (1303/82 Kirstenbosch). This very striking form was collected in the Nkandla Forest in Zululand.

1 ***P. saccatus*** **'King Goodwill' originates from a single white-flowered plant found amongst a group of mauve plants in Ngoye Forest in Zululand (KwaZulu-Natal)**

2 ***P. saccatus*** **'Nkandla' has a very attractive hooded and lined erect upper lip. It is a soft, fast-growing shrub**

1

2

Plectranthus sanguineus

Plectranthus sanguineus Britten in Trans. Linn Soc. London, Bot. 4: 36 (1894). Type: Malawi, Alexander Whyte 46, (K).

Name: '*sanguineus*' = blood-red, referring to the red anthers.

Common names: mountain spurflower, berg spoorsalie

Description: Sparingly branched succulent shrublet to 45 cm. *Stems* terete, dark maroon-brown, swollen towards base. *Roots* fleshy. *Leaves* broadly obovate to subrotund, dark green, succulent, aromatic, 25–60 × 33–50 mm; gland dots yellow; base cuneate, tip rounded, crenate in the upper two-thirds (3–5 pairs of teeth). *Inflorescence* racemose, simple or with a pair of side branches 40–150 mm long; peduncle tomentose; verticillasters 10–12 mm apart, crowded, about 40–60 flowered; bracts triangular-ovate, boat-shaped, 3 × 1 mm, early deciduous. *Calyx* 2 mm long, 5-toothed, upper tooth larger, subulate, 1.5 mm. *Corolla* 12–14 mm, dull mauve, deflexed at the base and expanding to the throat; lower lip boat-shaped, 3 mm long, with long hairs on the lower surface; upper lip 4-lobed, the upper emarginate, 1.5 mm, with long hairs on the adaxial side; lateral lobes white; filaments mauve, fused to the base of the corolla tube, free for 2 mm; anthers white; pollen orange to red; style 7 mm long. Fruiting calyx 3 mm. Nutlets 1 mm in diameter, dark brown.

Flowering time: Autumn.

Diagnostic features: The reddish anthers and swollen roots immediately distinguish it.

Distribution: Only known from the top of Mount Mulanje in Malawi.

Habitat: Grassland among granite rocks. *P. sanguineus* shares its habitat with *P. crassus*. Rainfall is 1500–2000 mm a year and falls mainly during summer.

History: Described by the British botanist, James Britten (1846–1924) in 1894 from plants collected by Alexander Whyte (1834–1908). It was introduced to Kirstenbosch in 1991 (210/96).

Cultivation: Best grown as a pot plant, or in rockeries in full sun. It is easily propagated from cuttings and roots rapidly. Resistant to light frosts.

Plate 51

Plectranthus sanguineus

Origin: Mount Mulanje, Malawi (210/96)

Plectranthus spicatus

Plectranthus spicatus E. Mey. ex Benth. in E. Mey., Comm. 230(1837); Codd in Bothalia 11: 383 (1975); in F. S. A. 28, 4: 146 (1985). Type: Cape, 'Glen-filling', Drège 4731b in Herb. Benth. (K, lecto.; MO; P; S).

Name: Latin, '*spica*' = a spike, referring to the spike-bearing inflorescence.
Common names: spike spurflower, ritsblom spoorsalie
Synonyms: *Burnatastrum spicatum*; *Plectranthus subspicatus*

Description: An erect, branched, succulent herb, 100–700 mm tall; stems 4-angled, decumbent to ascending. *Leaves* 13–40 × 7–29 mm, elliptic to obovate, pubescent to glabrescent; glands red; margin entire to deeply dentate (1–4 pairs of teeth); base cuneate; apex acute. *Inflorescence* subspicate, 90–300 mm long; flowers in decussate dichasia, up to 9-flowered. *Corolla* 8–10 mm long, blue-purple to violet-purple; tube deflexed, expanding to throat. *Calyx* 3 mm long, elongating to 5 mm after flowering. *Nutlets* shiny, brown, 1 mm long.

Flowering time: Autumn into winter (March to June with a peak at the end of April).

Diagnostic features: *Plectranthus spicatus* is related to *P. cylindraceus* but the latter has a densely villous inflorescence with smaller flowers (4–5 mm long) and a strong aroma of camphor.

Distribution: Widespread, occurring from the Drakensberg foothills to the coast, from the Eastern Cape to Mpumalanga in the north.

Habitat: Dry to semi-arid savannah. The plants are found on rocky outcrops in full sun or light shade. Rainfall, which occurs mainly in summer, is 400–750 mm a year, and frost is light or absent.

History: *P. spicatus* was discovered in 1837 by Johann Drège in the Eastern Cape. It was introduced to Kirstenbosch in 1935 by Jacob Gerstner (3313/35).

Cultivation: This is an attractive succulent suitable for areas where frost is not severe, and will suit succulent thicket or bushveld gardens. It also thrives as a pot plant. It propagates readily from seed, germinating in 21–28 days, or from cuttings taken in spring. Young plants should be spaced 200–300 mm apart in light shade or full sun.

'Nelspruit'. A form from the granite domes of the Nelspruit region in Mpumalanga with distinctly concave boat-shaped leaves which have entire margins (364/76 Kirstenbosch).

'Ubombo Mountain'. An attractive plant with small, succulent leaves with deeply dissected dark purple margins (1–2 pairs of teeth). The inflorescence is short and compact (758/77 Kirstenbosch).

Plate 52

Plectranthus spicatus

Origin: Near Ubombo on the Lebombo Mountains (758/77)

Plectranthus strigosus

Plectranthus strigosus Benth. in E. Mey. Comm. 229 (1837); Codd in Bothalia 11, 4: 409 (1975); in F. S. A. 28, 4: 158 (1985). Type: Cape Province. – 'Olifantshoek Forest' (Alexandria), Ecklon (K, lecto.).

Name: Latin, '*strigosus*', with short, dense bristles, referring to the stems and leaves.
Common names: dwarf spurflower, dwerg spoorsalie
Synonyms: *Plectranthus kuntzianus*; *P. parviflorus*

Description: Procumbent to decumbent, mat-forming, succulent herb, stems 30–100 mm long. *Branches* 4-angled, rusty strigose. *Leaves* 15–35 × 15–35 mm, ovate to broadly ovate, strigose to sparingly strigose, glands red; base truncate to cuneate, apex rounded or acute; margin shallowly crenate. *Inflorescence* terminal, racemose, 30–120 mm long, often with a pair of side branches; verticillasters up to 6-flowered. *Calyx* 3 mm, enlarging to 5–7 mm after flowering. *Corolla* 10–12 mm long, white, saccate at the base and constricted near the throat. *Nutlets* 1 mm, light to dark brown.

Flowering time: Spring to autumn (September–May, with a peak in April).

Diagnostic features: *P. strigosus* is highly variable, probably due to the dissected nature of its habitat which has led to the evolution of distinctive local forms. It closely resembles *P. purpuratus* but is distinguished by its rusty, strigose hairs. The corolla is also constricted near the throat, not in the middle as in *P. purpuratus*. The leaves of *P. purpuratus* are usually dark purple on the undersurface and covered in dense grey velvety hairs.

Distribution: Found only in the Eastern Cape.

Habitat: Plants occur on well-wooded, frost-free rocky ridges in kloofs and forested river valleys. Soils are well drained and humus-rich. Rainfall averages between 700 and 1250 per annum, falling mainly in summer in the north and throughout the year in the south of its range.

History: It was first collected in 1813 by William Burchell in the Bathurst district, and was introduced to Kirstenbosch in 1974 (151/74).

Cultivation: *P. strigosus* is best for subtropical gardens and ideal in a shady rockery where its succulent nature will enable it to survive dry periods. It will thrive in hanging baskets and containers, but will need to be raised afresh each year. Cuttings can be planted *in situ* during spring (about 150 mm apart). Rooting is fast and given ample moisture, compost and fertilizer, a closed cover soon forms and flowers will be produced in the same year. It needs to be pruned after flowering and before spring.

Plate 53

Plectranthus strigosus

Origin: Eastern Cape, near East London

Plate 54

Plectranthus stylesii

**Origin: Mzamba River,
Eastern Cape (494/06)**

Plectranthus stylesii

Plectranthus stylesii Edwards in Bothalia 35, 2: 149–52 (2005).

Name: After David Styles, a botanical explorer from Durban.
Common names: Msikaba spurflower, Msikaba spoorsalie

Description: Perennial, decumbent, faintly aromatic branched herb to 120 mm, roots shallow, fibrous. *Stems* semi-succulent, herbaceous, obscurely 4-angled, becoming subterete; surface densely strigose; hairs 0.4 mm and smaller, but nodes with longer hairs (0.5 mm). *Leaves* ovate to broadly ovate to almost subrotund, (35) 50–80 (–90) × (20–)30–40(–45) mm; apex acute; base cuneate; occasionally slightly decurrent on petiole, semi-succulent; abaxial surface reticulate-veined, sparingly strigose, becoming less so with age, covered with slightly sunken reddish gland dots; margin shallowly serrate with 4–5 pairs of teeth (occasionally with smaller secondary teeth). *Inflorescence* a short terminal raceme, 100–220 mm long, sometimes with a pair of side branches at the base; rachis sparsely strigose, bearing scattered sessile reddish gland dots; bracts lanceolate 8 × 2 mm. *Flowers* in sessile, 1–3-flowered cymes forming 2–6-flowered verticillasters, the latter 10–16 (–18) mm apart; pedicels 4.5–5 mm long, finely strigose. *Calyx* up to 5 mm long, lengthening to up to 7–8 mm in fruit. *Corolla* 2-lipped, white; lower portion of tube straight, 7 mm long, laterally compressed, expanding to 4 mm deep (3 mm from base), narrowing to 2.2–2.5 mm at the throat; sparingly covered with translucent gland-tipped hairs and red gland dots; upper lip 4-lobed, 5–6 mm high; the upper two larger lobes bent backwards, lateral lobes 3 mm high bent forwards. Lower lip horizontal, cymbiform, 5 mm long, its margins infolded; densely covered with red gland dots on lower surface. *Nutlets* rounded, 1.3–1.5 mm, brown, smooth.

Flowering time: Spring to autumn (September–May, with a peak in April).

Diagnostic features: It differs from *P. strigosus* mainly in its decumbent growth and larger shiny broadly ovate to almost subrotund leaves which are sparingly strigose, with hairs less than 0.4 mm long, and its larger corolla tube in which the upper lip lobes are bent backwards.

Distribution: From the Msikaba River in the south to the Mtamvuna in the north.

Habitat: Moist gorges in subtropical forest where annual rainfall is 1500–2000 mm, falling mainly between spring and autumn. The habitat is shared with plants such as *P. hilliardiae* subsp. *hilliardiae*, *P. saccatus* subsp. *pondoensis*, *Crassula streyi*, *C. orbicularis* and *Streptocarpus* spp. Trees and shrubs in the area include *Anastrabe integerrima*, *Apodytes abbottii*, *Ekebergia pterophylla*, *Cryptocarya latifolia* and *Lydenburgia abbottii*.

History: Although it has only recently been named *P. stylesii*, the plant itself has been known since the late seventies, and was until recently regarded as part of the *P. strigosus* complex, confined to the forested sandstone gorges of the Transkei region of the Eastern Cape. Two features – its decumbent growth, and the bending backwards of the upper lips of the corolla – show that it clearly belongs to a different entity. The first collection was made by the author and Geoff Nichols on 28 April 1979 in the Mtamvuna River (EvJ 3892, Compton Herbarium). During a recent expedition to the Msikaba River gorge, David Styles discovered a more succulent-leafed form (regarded as belonging to the same, but less succulent entities at Mtamvuna), which was described by Trevor Edwards in 2005 as a new species commemorating its finder.

Cultivation: It is best suited for coastal subtropical gardens and would thrive in a shady rockery where its succulent nature will help it survive dry periods. It can also be grown in a container. It is not a vigorous grower. Cuttings can be planted *in situ* during spring (about 150 mm apart). They will root quickly and with enough water, compost and food, will form a closed cover and flower the same year. Pruning is best done after flowering and before spring.

'Msikaba'. This selection from David Styles is markedly more succulent.

'Mtamvuna'. This has shiny leaves and is a less succulent form (406/89 Kirstenbosch).

Plectranthus swynnertonii

Woodbush Forest near Tzaneen, the habitat of *Plectranthus swynnertonii*. This species is normally white-flowered, but a pink form has recently been introduced from Woodbush

Plectranthus swynnertonii S. Moore in J. Linn. Soc., Bot. 40: 176 (1911); Codd in Bothalia 11: 422 (1975); in F. S. A. 28, 4: 166 (1985). Type: Zimbabwe, Chirinda Forest, Swynnerton 337 (K).

Name: After Charles Francis Massy Swynnerton (1877–1938), the English naturalist who first collected the plant.

Common names: cress spurflower, Debengeni spoorsalie

Description: Decumbent herb, up to 600 mm high; stems with a circlet of purple hairs at the nodes and dense purple hairs at the apex. *Leaves* membranous, 30–90 × 35–75 mm, broadly ovate, sparingly strigose or glabrous; glands flattened, yellow to honey-coloured; base cordate or truncate; apex acute; margin deeply serrate-dentate. *Inflorescence* a terminal raceme, 50–150 mm long; verticils 2 to 6-flowered. *Corolla* white to pale pink, 10–12 mm long, saccate at the base and narrowing towards the throat. *Calyx* 5 mm, elongating to 8 mm after flowering. *Nutlets* 1.75 mm, dark brown.

Flowering time: Summer into winter (November to June with a peak during April and, in mild climates, sporadically throughout the year).

Diagnostic features: The membranous, deeply dissected leaves and small saccate corolla are distinctive of *P. swynnertonii*.

Distribution: The species is recorded from Mpumalanga, the Chimanimani mountains of eastern Zimbabwe and Mount Mulanje in southern Malawi.

Habitat: Moist afromontane forest where it occurs as a groundcover, often along stream banks. Annual rainfall is between 1000 and 1750 mm and frost is absent.

History: It was first collected in Chirinda Forest in eastern Zimbabwe. It was introduced to Kirstenbosch in 1977 (497/77).

Cultivation: *P. swynnertonii* thrives as a groundcover in moist shady forest gardens where frost is not severe. The plants grow to 300–600 mm and are easily propagated by cuttings.

'Woodbush'. This selection has light pink flowers.

Plate 55

Plectranthus swynnertonii

Origin: Near Tzaneen, Northern Province (497/77)

Plectranthus tetensis

Plectranthus tetensis (Bak.) Agnew in Upland Kenya Wild Flowers 635 (1974); Codd in Bothalia 11: 390 (1975); in F. S. A. 28, 4: 148 (1985). Type: Mozambique, near Tete, Kirk s.n.(K, holo).

Name: '*tetensis*', from Tete Province in Mozambique were the plant was first described.

Common names: Tete spurflower, bushveld spurflower, Tete spoorsalie, bosveld spoorsalie

Synonyms: *Coleus decumbens*; *C. tetensis*; *C. vagatus*; *Plectranthus vagatus*

Plate 56

Plectranthus tetensis

Origin: Near Hoedspruit, Mpumalanga (1067/77)

Description: Weak perennial or annual succulent herb, procumbent; stems to 600 mm long, 4-angled, rooting at the nodes. Short side branches often drop at the end of autumn. *Leaves* 15–25 × 12–20 mm, ovate to elliptic, pubescent, glands pink; margin dentate (4–5 pairs of teeth); base cuneate; apex acute. *Inflorescence* racemose 80–220 mm long, flowers in 8 to 12-flowered verticillasters. *Corolla* up to 16 mm long, purple, geniculate, expanded to the throat. *Bracts* conspicuous, persistent, ovate, fleshy; apex rounded. *Calyx* 4–5 mm, elongating to 5–6 mm after flowering. *Nutlets* dark brown, 1 mm long.

Flowering time: Autumn into winter (March to the end of June with a peak in April).

Diagnostic features: Its procumbent habit and large, persistent bracts make *P. tetensis* distinctive. It is sometimes confused with *P. neochilus*, but the latter has decumbent stems and large membranaceous, deciduous bracts, which are apically acute. Its brittle side branches which drop at the end of the season are unique. When flowering is over, the mother plant dies in the way of an annual, but some of the short side branches drop to the ground to root during the spring, continuing the cycle. The plant is thus a type of 'vegetative annual'.

Distribution: Widespread in eastern Africa, from Kenya to northern KwaZulu-Natal.

Habitat: Plants grow in full sun or light shade in brackish, sandy, well-drained soil. Rainfall ranges between 300–700 mm per annum and frost is light or absent.

History: The species was described in 1900 by Edmund G. Baker (1864–1949) from plants collected by a Scot, John Kirk (1878–1962), a military man by profession who did some plant-collecting in eastern Africa. It was introduced to Kirstenbosch in 1977 (1067/77). It was originally named *Coleus tetensis* and transferred to *Plectranthus* by Leslie Codd in 1974.

Cultivation: *P. tetensis* is best suited for dry bushveld gardens in full sun or light shade. It grows rapidly and its brittle branches root when broken. It thrives in rockeries in well-drained soils but the plants are prone to mildew in moist areas.

Plectranthus tetragonus

Plectranthus tetragonus Gürke in Bot. Jb. 19: 109 (1894); Codd in F. S. A. 28, 4: 141 (1985). Type: Tanzania, Usambara, Mashena, Carl 3573.

Name: Greek, '*tetragonus*' = four-angled, referring to the square stems.

Common names: annual spurflower, opslag spoorsalie

Synonyms: *Coleus melanocarpus*; *Englerastrum tetragonus*; *Plectranthus biflorus*; *P. melanocarpus*

Description: Erect annual herb up to 300–600 mm. *Stems* 4-angled with spreading bristles. *Leaves* 30–100 × 20–60 mm; blade glabrescent to oblong-ovate; base obtuse or cuneate; apex acute. *Inflorescence* racemose, 40–120 mm long, terminal. *Flowers* pedicellate, solitary or paired. *Corolla* 4–6 mm long, yellow, geniculate and expanding to the throat. *Calyx* 2 mm long, elongating up to 10 mm. *Nutlets* 1 × 0.7 mm long, shiny black.

Flowering time: Autumn (March to April).

Diagnostic features: This species shares yellow flowers with *P. esculentus*, which is easily distinguished by its trailing stems and perennial tuberous roots; *P. tetragonus* is an erect annual.

Distribution: Widespread from the North-West province through Swaziland, Botswana, Zimbabwe, Zambia, Mozambique and Tanzania.

Habitat: Dry savannah in full sun or semi-shade, usually in sandy soils. Rainfall averages between 600 and 1250 mm per annum and falls mainly during summer. Frost is light or absent.

History: *P. tetragonus* was first described by Gürke in 1894 from plants discovered by the German plant collector Carl Holst at Mashena, Usambara in Tanzania.

Cultivation: Plants are easily propagated from seeds, but are of little ornamental merit. Can be used as an annual in dry bushveld gardens.

Plate 57

Plectranthus tetragonus

Origin: North of Thabazimbi, Limpopo Province

Plectranthus unguentarius

Plectranthus unguentarius Codd in Bothalia 11, 4: 387–388 (1975). Type: Kaokoveld, 17 km south of Kaoko-Otavi on road to Sesfontein, 21 April 1957. De Winter & Leistner 5595 (PRE, holo).

Name: '*unguentarius*' = perfumed.
Common names: Kaoko spurflower, Kaoko spoorsalie

Description: Erect, sparingly branched, deciduous, pleasantly aromatic shrub to 1–3 m. *Branches* obscurely 4-angled at the base, surface hairy. *Roots* tuberous. *Leaves* fleshy, broadly obovate to subrotund, 40–60 × 40–60 mm; surface hairy and with reddish-brown gland dots; lower surface reticulately veined; margin crenate; petiole to 15 mm long, base cuneate; apex rounded. *Inflorescence* a terminal raceme to 350 mm long, bearing densely glomerate verticils 10–30 mm in diameter; rachis densely woolly-tomentose; bracts broadly ovate 3 × 2.5 mm, acute. *Corolla* 10–15 mm long, white to greenish-white; tube slightly bent, 1 mm deep at the base and expanding to 2 mm; upper lip erect, hooded; lower lip 10 mm long; stamens 10 mm long. *Calyx* 2–3 mm, elongating 4–5 mm after flowering. *Nutlets* not seen.

***Plectranthus unguentarius*, here found growing on a granite outcrop in the dry northern Kaokoveld. It is a fast grower with aromatic leaves, ideal for dry bushveld gardens**

Flowering time: Autumn (March–April).

Diagnostic features: *P. unguentarius* is characterized by its erect stems to 2 m, its deciduous, semi-succulent obovate leaves which are pleasantly aromatic, its dense glomerate verticils, and its greenish-white to white hooded corollas. It is related to *P. porcatus* which has rough sandpaper-like bark, and to *P. xerophilus* which has dark purple flowers.

Distribution: Endemic to the Kaokoveld from Otjihipa in the north to Kaoko-Otavi in the south.

Habitat: Arid savannah with *Colophospermum mopane* dominant. Plants occur on well-drained soil in full sun on rocky dolomite or granite outcrops. Other plants include *Adenium boehmianum*, *Boschia albitrunca*, *Catophractes alexandri*, *Euphorbia subsalsa*, *Pachypodium lealii*, *Sesamothamnus leistneri* and *Terminalia prunoides*. Rain occurs mainly in summer and averages 400–700 mm per annum. Frost is absent from the habitat.

History: It was first collected by Bernard de Winter and Otto Leistner south of Kaoko-Otavi on their expedition to Kaokoveld in 1957, but not introduced to cultivation. This species took the author many years and many expeditions to relocate, but live material finally arrived at Kirstenbosch in 2005. See also pp. 23–5.

Cultivation: *P. unguentarius* makes an ideal, very drought-resistant, fast-growing shrub for dry bushveld gardens. It is best grown in full sun or light shade in well-drained soil. It is easily grown from cuttings which should flower the same season if planted in spring or summer; these can be set about 800 mm apart. By May the plants lose their leaves, becoming completely deciduous in winter.

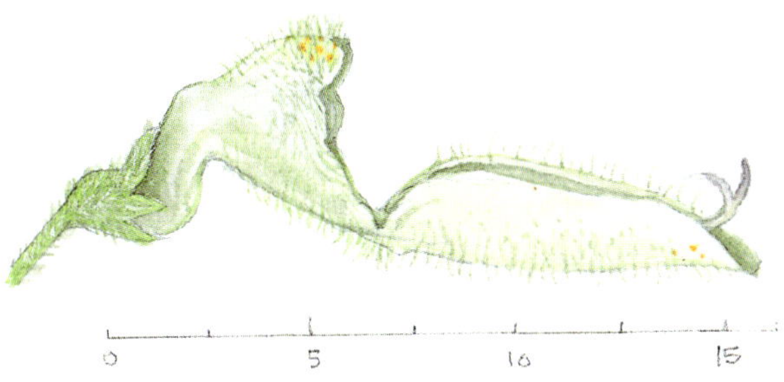

Plate 58

Plectranthus unguentarius

Origin: Kaokoveld, northern Namibia (124/05)

Plate 59

Plectranthus venteri

Origin: Sekukuniland (155/96)

Plectranthus venteri

Plectranthus venteri Van Jaarsv. & Hankey in Aloe 34 (1–2): 40–41, ills., 1997. Type: RSA, Limpopo Prov. Venter 13626 (PRE).

Name: Named after Fanie Venter, a well-known botanist, ecologist and co-author of a book on trees, who discovered the plants in Sekukuniland.
Common names: Sekukuni spurflower, Sekoekoenie spoorsalie

Description: Erect to decumbent pleasantly aromatic, multi-stemmed hairy shrub or subshrub to 700 mm high and 1 m in diameter. *Roots* slightly fleshy. *Stems* terete, mostly erect, rooting when decumbent and occasionally bearing stilt-roots. *Leaf blade* irregularly 5-lobed, triangular to broadly ovate in outline, 12–27 × 12–30 mm; base truncate; apex obtuse to acute. *Inflorescence* a simple erect raceme 180–280 mm long, occasionally with a pair of side branches at the base. *Flowers* in 6–8-flowered cymes forming 12–16-flowered verticillasters 3–12 mm apart. *Calyx* 3 mm long at flowering, enlarging to 6 mm. *Corolla* 8–10 mm long, violet-purple, tube slightly bent, 4.5 mm long. Stamens 4 (1 pair slightly longer), fused at the base, free for about 2 mm; pollen orange. Stigma bifid, slightly longer than stamens. *Nutlets* 1.5 mm in diameter, black, flattened, circular.

Flowering time: Autumn (March to May).

Diagnostic features: *P. venteri* is a distinct, clear-cut species not likely to be confused with any other South African plectranthus. It is at once distinguished by its uniquely lobed leaves, which bear shortly stalked gland dots and have a distinct pleasantly aromatic aroma. It can be distinguished further by its erect stems, short axillary branches, and violet-purple corolla. The upper lip of the calyx spreads horizontally during flowering, becoming erect during fruiting.

Distribution: A hilly area of Sekukuniland in Limpopo Province.

Habitat: Plants grow among norite boulders in savannah (bushveld), mostly in full sun but also in the shade of rocks, and usually in shallow gravelly soil which dries out between rainy spells.

History: *P. venteri* was only named in 1997. It was discovered by Fanie Venter and was first introduced to cultivation in 1996 (Kirstenbosch 155/96).

Cultivation: This long-lived perennial with a unique and pleasant aroma flourishes in cultivation, and is ideal for bushveld gardens. It is very drought-tolerant and would do well in a rockery or as a container plant. Its highly aromatic leaves contain etheric oils which may well have economic potential. It thrives in many areas, even in the Western Cape.

Plectranthus venteri **in its native habitat growing on norite boulders in Sekukuniland (Limpopo Province). Its lobed leaves and fresh aroma are unique in the genus**

Plectranthus verticillatus

Plectranthus verticillatus (L.f.) Druce in Rep. Bot. Exch. Cl. Brit. Isles (1916); Codd in Bothalia 11, 4: 407 (1975); in F. S. A. 28: 4: 157 (1985). Type: Cape Province. – 'India Montin', (Linn 749.4 holo.).

Name: Latin '*verticillatus*', whorled, referring to the arrangement of flowers on the inflorescence.
Common names: gossip spurflower, money plant, skindersalie
Synonyms: *P. nummularius*, *P. thunbergii*

Description: Procumbent, mat-forming, succulent herb; stems shortly strigose, 30-500 mm long. *Leaves* 14–48 × 12–45 mm, ovate to rotund, glabrescent to occasionally pubescent, glossy, glands red; base broadly cuneate to truncate; petiole 15–30 mm; apex acute to rounded; margin shallowly crenate. *Inflorescence* terminal, racemose, 50–260 mm long, often with a pair of side branches at the base, verticillasters up to 6-flowered. *Calyx* 3 mm long, enlarging to 5 mm. *Corolla* 10–25 mm long, white, often speckled; tube 4–16 mm long, straight, laterally compressed, 2–3 mm broad, almost linear; base slightly saccate. *Nutlets* 1 mm, brownish.

Flowering time: Autumn to winter (March to July).

Diagnostic features: This species is closely related to *P. strigosus* and *P. purpuratus* but is easily distinguished by its corolla which is larger and more or less parallel-sided, and by its longer stamens of 5–7 mm. The corolla tubes of the other two species are shorter and constricted in the middle, and their stamens are 3–5 mm long.

Distribution: From Knysna to Limpopo Province.

Habitat: Wooded, frost-free rocky ridges in kloofs and forested river valleys in well-drained soil rich in humus. Rainfall is between 700 and 1750 mm per annum.

History: It was first collected by Carl Thunberg in the late eighteenth century, and was described by the younger Linnaeus in 1781 as *Ocymum verticillatum*. It came to Kirstenbosch in 1915 (170/15). This species is extensively grown in Europe and America.

Cultivation: With its attractive, often glossy leaves, procumbent form and soil-binding qualities this is a very popular groundcover where frost is not severe. Many forms have purple undersides, and this coloration is accentuated during periods of stress. The common names 'Gossip' and 'Money plant' refer to its quick growth rate and its almost rounded, shiny leaves. Its succulence enables it to survive dry periods. It makes a useful container plant, though will need annual renewal. Cuttings can be taken at the end of winter and planted *in situ* about 300 mm apart. It should be pruned after flowering. Because of its wide distribution this is a highly variable species with many local forms.

'Barberton'. An attractive form with dark reddish purple undersides (1302/82 Kirstenbosch).

'Blyde'. Large, glossy, broadly ovate leaves (55 × 62 mm), with 6–7 pairs of teeth. This form is grown extensively in the United States (440/84 Kirstenbosch).

'Eureka'. An attractive variant with small hairy leaves with purple undersides (426/84 Kirstenbosch). It originates from the former gold-rush town 'Eureka City', near Barberton.

'Freckles'. A cultivar with spotted variegated leaves, distributed by Glasshouse Works Nursery in Ohio (USA) (1369/82).

'Gossip'. A variegated cultivar distributed by Glasshouse Works Nursery in Ohio (USA) (1368/82).

'Malelaan'. This form is possibly a natural hybrid between *P. verticillatus* and *P. fruticosus*. It was introduced from the Krokodilpoort Mountains near Malelaan by Koos Nel of the Lowveld Botanical Gardens. It has pink flowers and large leaves (70–80 mm) with 4 to 5 pairs of marginal teeth (202/83 Kirstenbosch).

'Money Maker'. A procumbent form with rounded leaves from the Kei River (815/2002 Kirstenbosch).

'Pink Surprise'. A mat-forming plant with green leaves and surprising pink flowers. This selection originates from the sheer cliff-faces at Geelhoutboskloof in the Baviaanskloof. It makes an ideal groundcover and is good in a hanging basket (697/91). See p. 7.

'Ubombo'. This attractive form with flaccid stems and wavy leaves with pronounced veins on the upper surface is widespread on the Lebombo Mountains, and excellent for hanging baskets and rockeries. Introduced to Kirstenbosch in 1977 (756/77 Kirstenbosch).

Plate 60

Plectranthus verticillatus

Origin: Near Barberton in Mpumalanga (1302/82)

Plectranthus woodii

Plectranthus woodii Gürke in Bot. Jahrb. 26: 76 (1889). Type: Biggarsberg, Rehmann 7092 (Z, holo.).

Name: After John Medley Wood (1827–1915), an English botanist and explorer in KwaZulu-Natal.
Common names: Wood's spurflower, vetwortel spoorsalie, Wood spoorsalie
Synonyms: *Plectranthus draconis*; *P. hadiensis* var. *woodii*; *P. zatarhendi* var. *woodii*

Plate 61
Plectranthus woodii
Origin: Ex hort.

Description: Perennial herb up to 600 mm, rootstock tuberous, branches spreading. *Leaves* 20–60 × 15–50 mm, ovate to broadly ovate, puberulous to tomentose; base cuneate to subcordate; apex acute to rounded; margin crenate-dentate with 4–7 pairs of teeth. *Inflorescence* terminal, racemose up to 350 mm long, occasionally with a pair of side branches; flowers in 4–8-flowered verticillasters, 10–30 mm apart. *Corolla* 8–12 mm long, mauve to white; tube deflexed, widening towards the throat. *Calyx* 2–3 mm, enlarging to 5–6 mm after flowering. *Nutlets* 1.2 mm, brown.

Flowering time: Autumn into winter (March to June with a peak in April).

Diagnostic features: *Plectranthus woodii* is closely related to *P. hadiensis*, *P. madagascariensis* and *P. grandidentatus*, but its tuberous roots are distinctive. It was formerly considered a variety of *P. hadiensis*, but on the grounds of its tuberous roots it is here regarded as a separate entity.

Distribution: From the Eastern Cape and KwaZulu-Natal to Mpumalanga in the north.

Habitat: It is a species of savannah almost always growing in humus pockets on rocky outcrops. Plants from the Mpumalanga highveld are subject to seasonal frost and veld fires and re-sprout annually from their persistent tuberous rootstocks. The soil is well drained and humus-rich. Rainfall ranges from 600 to 1000 mm per annum.

History: *P. woodii* was first collected by John Medley Wood at Ipolweni, KwaZulu-Natal. It was introduced to Kirstenbosch in 1978 (258/78).

Cultivation: This white-flowered, tuberous-rooted species can be grown in full sun, and would suit a rockery in a grassland or sunny highveld garden. It is easily propagated from cuttings and grows rapidly; young plants can be set out 300 mm apart, and should flower the same year. It is best to prune after flowering.

Plectranthus xerophilus

Plectranthus xerophilus Codd in Bothalia 11: 282 (1974); Codd in F. S. A. 28, 4: 142 (1985). Type: Transvaal, Lydenburg District, near Marone, Codd & Dyer 7729 (PRE, holo.).

Name: Greek, '*xerophilus*' = dry-loving, referring to its dry habitat.
Common names: Steelpoort spurflower, Steelpoort spoorsalie

Description: Erect, sparingly branched, deciduous shrub, 1–2 m., bearing rounded, not square branches. *Roots* horizontal, tuberous. *Leaves* 35–90 × 25–70 mm, ovate to elliptic, leathery, glandular-tomentose; glands russet; margin crenate; base cuneate or truncate; apex obtuse. *Inflorescence* paniculate, 350–680 mm long, flowers in 6 to 25-flowered verticillasters. *Corolla* 13–14 mm long, purple; tube deflexed, expanding to throat; upper lip distinctly hooded. *Calyx* 2.5 mm, elongating to 4 mm after flowering. *Nutlets* flattened, 0.75 mm long.

Flowering time: Autumn into winter (March to June with a peak in May).

Diagnostic features: *P. xerophilus* is characterized by its horizontal tuberous roots, terete branches and hooded corolla. Initially the stigma is trapped in a groove of the upper lip but is later released, the significance of which is unknown.

Distribution: Endemic to the mountains of Mpumalanga and Limpopo Province where it seems to be confined to dolomite outcrops.

Habitat: Plants occur on sandy soils in savannah and are frequently exposed to veld fires. Rain occurs mainly in summer and averages 400–700 mm per annum. Frost is light or absent.

History: It was first collected by Inez Verdoorn and Amelia Obermeyer at Malipsdrift (Limpopo Province) in 1935, and was introduced to Kirstenbosch in 1977 (671/77).

Cultivation: *P. xerophilus* is a drought-resistant shrub best grown in warm dry bushveld and subtropical gardens in the northern and eastern parts of South Africa. Prune after flowering. It is easy to propagate from cuttings, and these when rooted should be planted about 800 mm apart in full sun or light shade in well-drained, composted soil.

Plate 62

Plectranthus xerophilus

Origin: Abel Erasmus Pass, Mpumalanga (671/77)

Plectranthus zuluensis

Plectranthus zuluensis T. Cooke in Kew Bull. (1909); Codd in Bothalia 11, 4: 424 (1975); in F. S. A. 28, 4: 167 (1985). Type: Natal, Gerrard 1675 (K, holo.).

Name: '*zuluensis*', from Zululand in KwaZulu-Natal.
Common names: Zulu spurflower, Zoeloe spoorsalie

Description: Erect or rarely decumbent herb or shrub, up to 2 m, stems shortly pubescent. *Leaves* 35–110 × 25–80 mm, ovate to broadly ovate, shortly pubescent, glands colourless; base cuneate to truncate; apex acute to obtuse; margins crenate with 8–12 pairs of teeth. *Inflorescence* a terminal raceme 60–100 mm long, verticillasters 6-flowered. *Calyx* 3–6 mm enlarging to 8 mm after flowering. *Corolla* 8–16 mm long, blue-mauve (rarely white); saccate at the base narrowing to the throat. *Stamens* 4, the upper pair sterile. *Nutlets* 1.5 mm, dark brown.

'Oribi Gorge' is one of the best of the *P. zuluensis* selections. It is a fast-growing shrub flowering in summer and autumn, happiest in shade

Flowering time: Spring to winter (October to June with a peak during April. Also sporadically throughout the year except between July and September).

Diagnostic features: The reduction of the upper stamens to staminodes distinguishes *P. zuluensis* from other species.

Distribution: Widespread in subtropical forest from the northern part of the Eastern Cape through to KwaZulu-Natal and Swaziland in the north.

Habitat: Plants occur in forest and along forest margins, in hilly terrain or deep river gorges, often along stream banks. Soils are humus-rich. Rain falls in summer and averages 1000–1500 mm per annum; frost is light or absent.

History: It was first collected by William Gerrard, an English traveller and naturalist, but was only introduced to cultivation in 1934 by R. A. Dyer at the Botanical Research Institute where it was grown and illustrated for *Flowering Plants of Africa*. It came to Kirstenbosch in 1977 (810/77).

Cultivation: *P. zuluensis* is an attractive shrub best suited for warm subtropical forest gardens. It can be maintained as a shrubby groundcover but will need some pruning. In humid areas it can tolerate full sun. In the garden it flowers freely and copes with drought. A natural hybrid between it and *P. saccatus* found in the Ngoye Forest has been grown under the name 'Zulu Dance'.

'Oribi Gorge'. A very attractive shrub with light purple-blue flowers and velvety leaves (810/77 Kirstenbosch).

'Sky'. A low-growing shrub to 1 m with dark purple-blue flowers. It was introduced to Kirstenbosch in 1989 from the University of KwaZulu-Natal Botanical Gardens in Pietermaritzburg (Kirstenbosch 752/89), but its wild origin is unknown.

'Umgai'. A shrub up to 2 m with dark purplish blue flowers. It was introduced to the Durban Botanic Garden by Rudolf Strey (119/82 Kirstenbosch).

Plate 63

Plectranthus zuluensis

Origin: Oribi Gorge, KwaZulu-Natal (810/77)

Distribution maps

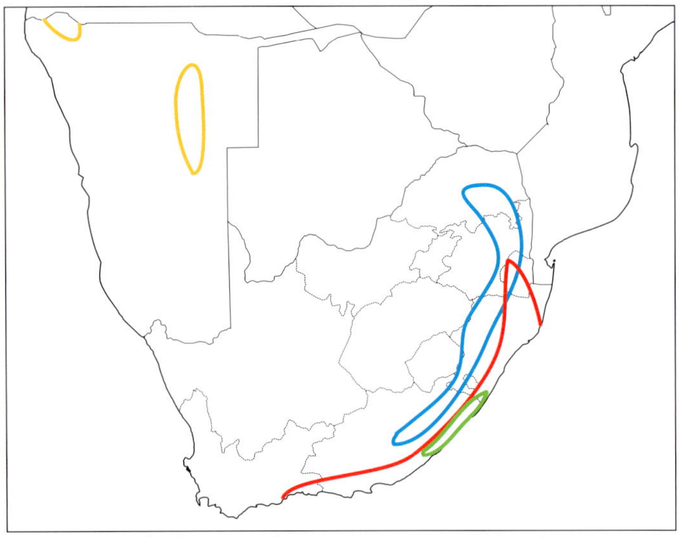

Distribution of subgenus *Plectranthus* section *Coleoides*:
■ P. aliciae ■ P. dinteri ■ P. grandidentatus
■ P. madagascariensis

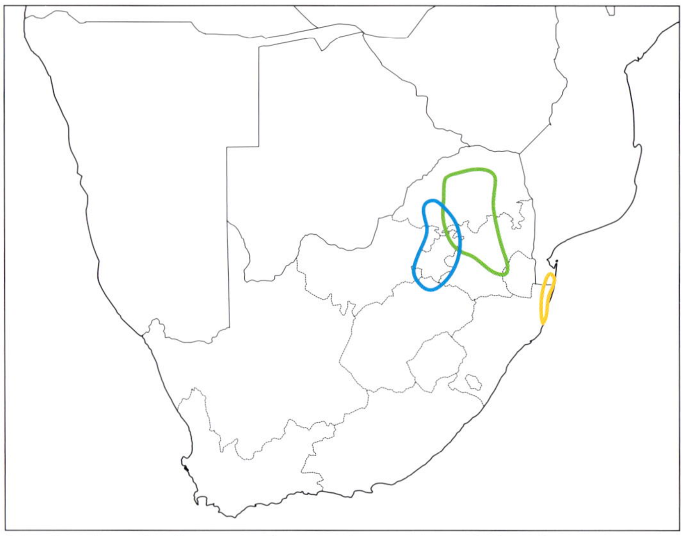

Distribution of subgenus *Plectranthus* section *Coleoides*:
■ P. mutabilis ■ P. psammophilus ■ P. ramosior

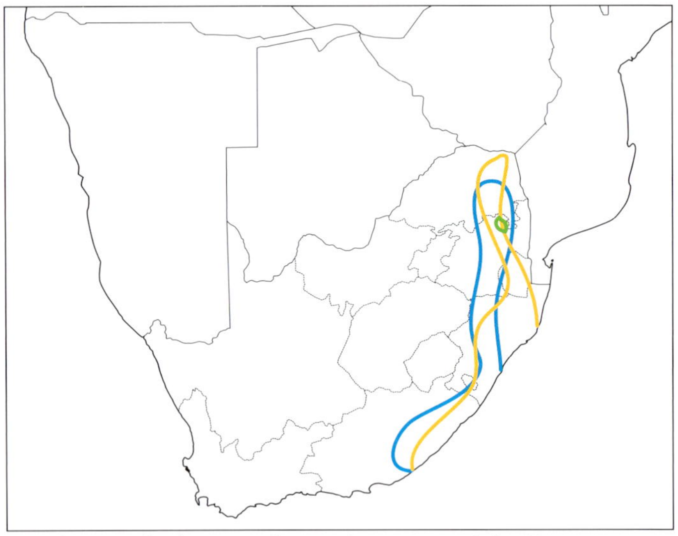

Distribution of subgenus *Plectranthus* section *Coleoides:*
■ P. venteri ■ P. hadiensis ■ P. woodii

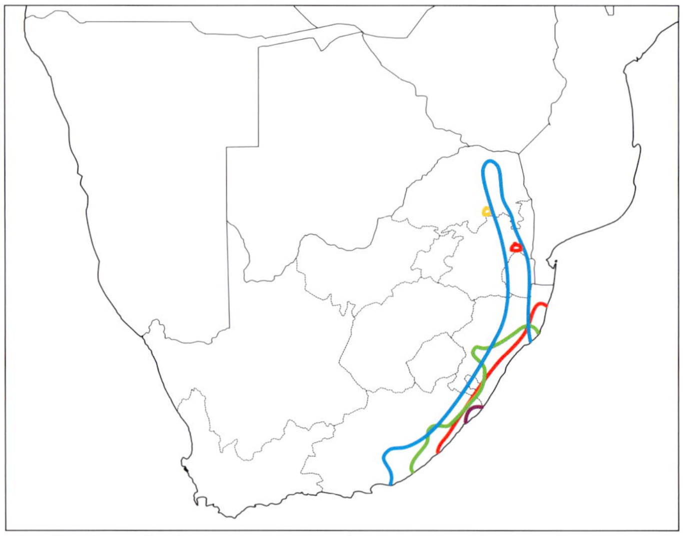

Distribution of subgenus *Plectranthus* section *Plectranthus:*
■ P. ambiguus ■ P. dolomiticus ■ P. laxiflorus
■ P. petiolaris ■ P. reflexus

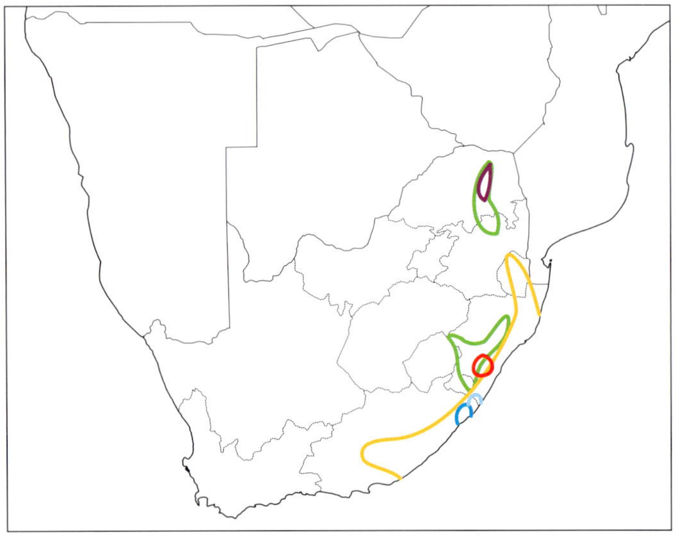

Distribution of subgenus *Plectranthus* section *Plectranthus:*
■ P. dolichopodus ■ P. ecklonii ■ P. hilliardiae ssp. australis
■ P. hilliardiae ssp. hilliardiae ■ P. rehmannii ■ P. swynnertonii

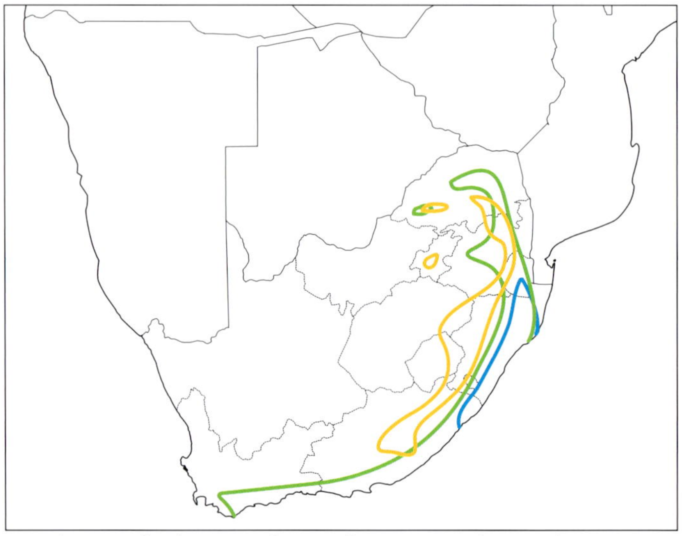

Distribution of subgenus *Plectranthus* section *Plectranthus:*
■ P. fruticosus ■ P. grallatus ■ P. zuluensis

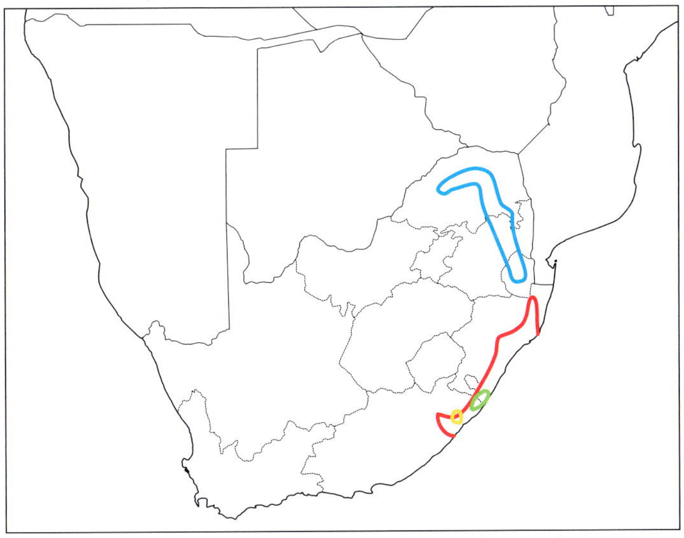

Distribution of subgenus *Plectranthus* section *Plectranthus:*
■ P. ernstii, P. saccatus ssp. pondoensis ■ P. mzimvubuensis
■ P. rubropunctatus ■ P. saccatus ssp. saccatus

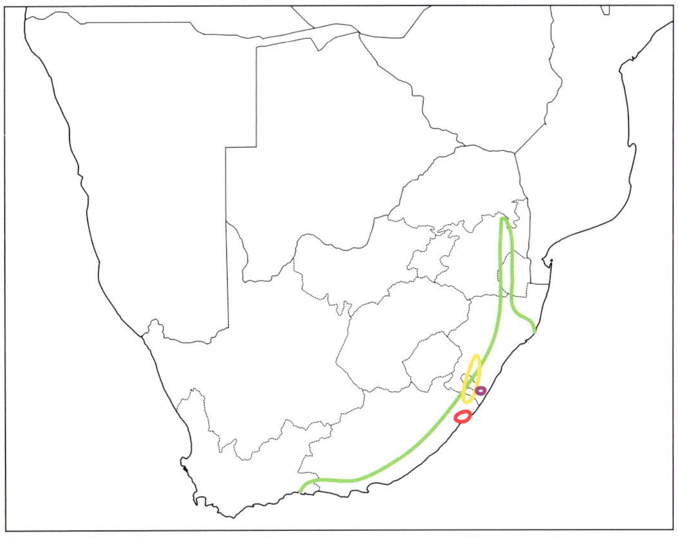

Distribution of subgenus *Plectranthus* section *Plectranthus:*
■ P. ciliatus ■ P. elegantulus ■ P. oertendahlii, P. oribiensis
■ P. malvinus, P. praetermissus

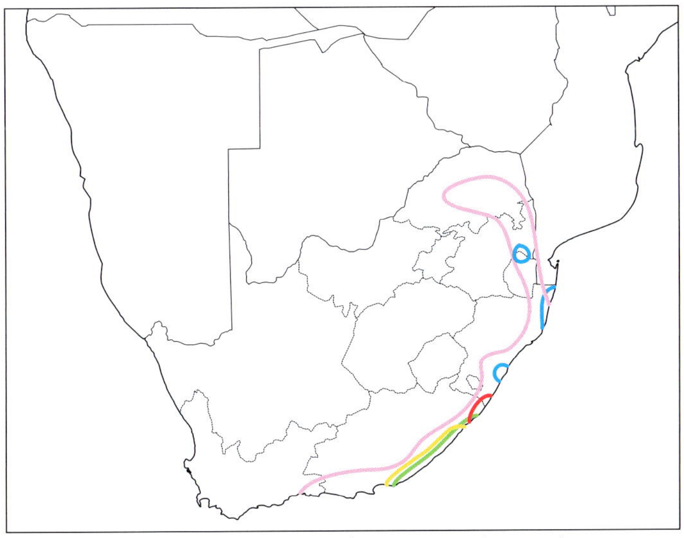

Distribution of subgenus *Plectranthus* section *Plectranthus:*
■ P. verticillatus ■ P. lucidus ■ P. strigosus ■ P. purpuratus ssp. purpuratus (bottom), ssp. montanus (top), ssp. tongaensis (middle) ■ P. stylesii

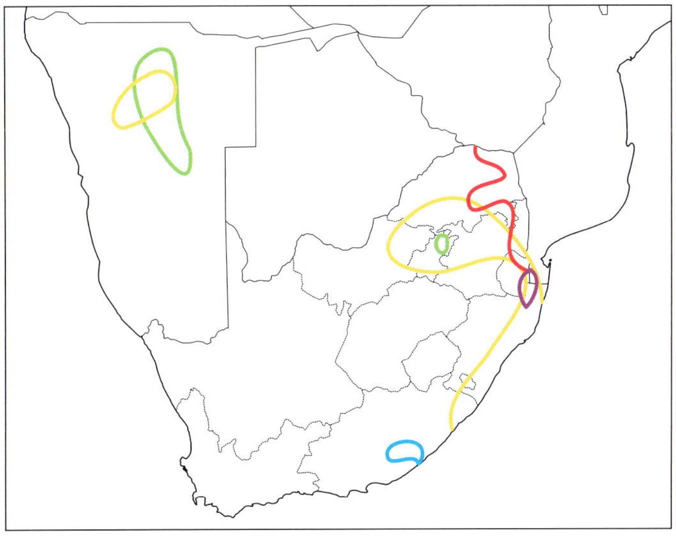

Distribution of subgenus *Calceolanthus:*
■ P. caninus ■ P. neochilus ■ P. pentheri ■ P. tetensis
and subgenus Coleus: ■ P. amboinicus

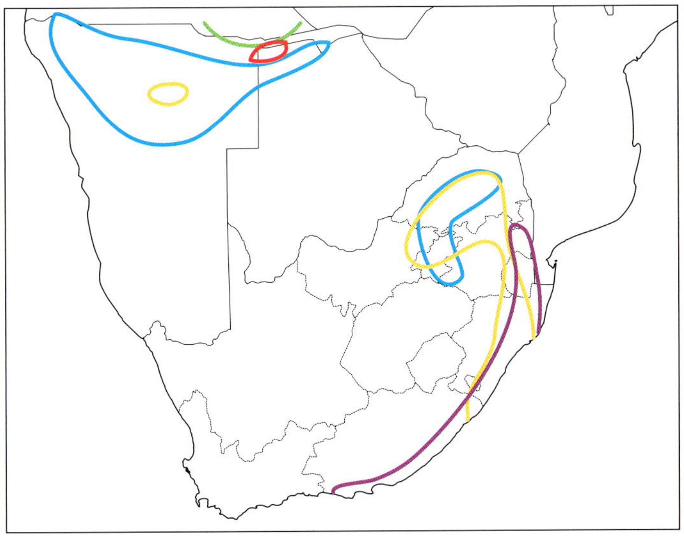

Distribution of subgenus *Burnatastrum:*
■ P. candelabriformis ■ P. cylindraceus
■ P. hereroensis ■ P. mirabilis ■ P. spicatus

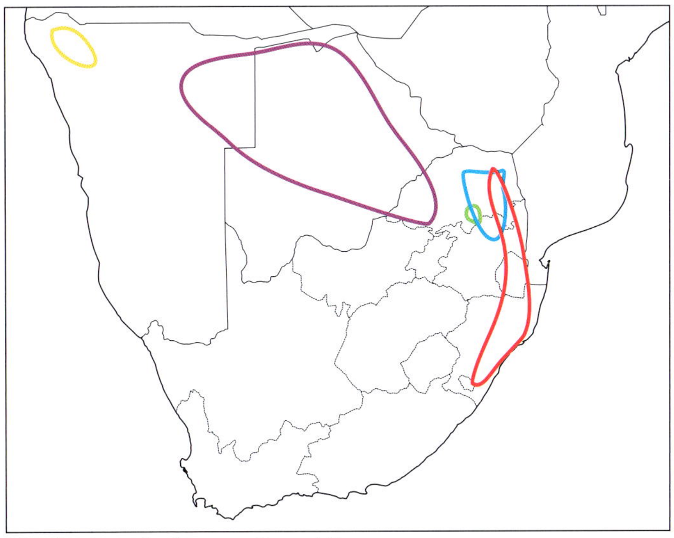

Distribution of subgenus *Xerophilus:*
■ P. porcatus ■ P. unguentarius ■ P. xerophilus
and subgenus Nodiflorus: ■ P. esculentus ■ P. tetragonus

Bibliography

——1996. P. ciliatus 'Easy Gold' – an Australian cultivar. *Plant Varieties Journal.* 9: 50–1.

——1998. Plectranthus oertendahlii. In: Danish pot plants. Pot Plant Catalogue, Gasa Odense – Blomster A.m.b.a., Lavsenvaenget 1, DK-5200 Odense V, Denmark.

——2000. Van der Voorts winnen noviteitenshow. *Vakblad voor de Bloemisterij.* 11: 7.

Acocks, J. P. H. 1975. *Veld Types of South Africa.* Botanical Survey Memoir No. 40. Botanical Research Institute, Pretoria.

Aiton, W. T. 1810. *Hortus Kewensis* 2nd ed., Vol. 2. London.

Andrews, H. C. 1804–15. *The Botanist's Repository.* Vols. 1–10. London.

Bailey, L. H. 1916. *The Standard Encyclopedia of Horticulture.* New York.

Bentham, G. 1832. *Labiatarum Genera et Species.* London.

Bentham, G. 1848. Labiatae. In Augustin Pyramus de Candolle, *Prodromus* 12: 55.

Bentham, G. & Hooker, Sir J. D. 1876. *Genera Plantarum* 2: 1175.

Blake, S.T. 1971. A Revision of *Plectranthus* (Labiatae) in Australasia. *Contributions from the Queensland Herbarium* no. 9, Queensland.

Bradlow, F. 1965. *Baron von Ludwig and the Ludwigsburg Garden.* Cape Town.

Brickell, C. 1979. Pruning, in *The Royal Horticultural Society's Encyclopedia of Practical Gardening.* Mitchell Beazley, London.

Briquet, J., 1897. Labiatae. *Naturlichen Pflanzenfamilie* 4, 3a: 348–63. Berlin.

Brits, G. J., Selchau, J. and van Deuren, G., 2001. Indigenous *Plectranthus* (Lamiaceae) from South Africa as new flowering pot plants. In: Strategies for new ornamentals: Proceedings of the Twentieth International Eucarpia Symposium, Section Ornamentals. *Acta Horticulturae* 552: 165–170.

Brown, Laurian. 1997. Spur of the moment (Plectranthus). *House and Leisure.* December: 126–9.

Codd, L. E. 1975. *Plectranthus* (Labiatae) and allied genera in Southern Africa. *Bothalia* 11, 4: 371–442.

Codd, L. E. 1985. Lamiaceae in *Flora of Southern Africa,* ed. O. A. Leistner, Vol. 28 Part 4. Botanical Research Institute, Pretoria.

Cowling R. M. & Proches S. 2005. Patterns and evolution of plant diversity in the Cape floristic region. *Biologiske Skrifter* 55: 273–88.

Desmond, Ray. 1994. *British and Irish Botanists and Horticulturists.* Natural History Museum, London.

Edwards, T. J. 2005. Two new Plectranthus species (Lamiaceae) and new distribution records from the Pondoland Centre for Plant Endemism, South Africa. *Bothalia* 35, 2: 149–52.

Eggli, U. (ed.). 2002. *Illustrated Handbook of Succulent Plants: Dicotyledons.* Springer-Verlag, Berlin.

Gunn, M., & Codd, L. E. 1981. *Botanical Exploration of Southern Africa.* A.A. Balkema, Cape Town.

Hessayon, D. G. 1980. *The House Plant Expert.* Hazel Watson Viney, Herts.

Heywood, V.H. (ed). 1978. *Flowering Plants of the World.* Oxford University Press.

Hitchcock A., 1990. A show of Plectranthus. *Veld and Flora.* March: 24–6.

Hitchcock A., 1991. Plectranthus in the landscape. *Parks and Grounds* 58: 8–13.

Hutchings, A. 1996. *Zulu Medicinal Plants, an inventory.* University of Natal Press, Pietermaritzburg.

Jessop, J. 1980. Australian Lamiaceae. *Australian Plants* Vol. 10 no. 83. New South Wales.

Lanjouw, J., & Stafleu, F. A. 1954. *Index Herbariorum,* parts 1 & 2. Utrecht, Netherlands.

Launert, E., 1968. Miscellaneous notes on Labiatae. *Mitteilungen aus der Botanischen Staats sammlung München* 7: 295–307.

L'Héritier, C. L. 1784. *Stirpes Novae.* Paris.

Liberty, H. B. & Bailey, E. Z. 1976. *Hortus Third.* Macmillan, New York.

Loudon, J. C. 1838. *Arboretum et Fruticetum Britannicum,* Vol. 3. London.

Miller, D. & Morgan, N., 2000. A new leaf. In: *The Garden: Journal of the Royal Horticultural Society.* 125 (11): 842–5.

Palmer, Eve, 1989. *Under the Olive.* Tafelberg, Cape Town.

Phillips, E. P. Miscellaneous contributions in *Flowering Plants of Africa.* Plate 294. Government Printer, Pretoria.

Potgieter, C. J., Edwards, T. J., Miller, R. M. & Van Staden, J. 1999. Pollination of seven *Plectranthus* spp. (Lamiaceae) in southern Natal, South Africa. *Plant Systematics and Evolution.* 218: 99–112. Springer-Verlag, Austria.

Retief, E. & Herman, P. P. J. 2003. Lamiaceae. In G. Germishuizen & N. L. Meyer (eds), *Plants of Southern Africa: an annotated checklist. Strelitzia* 14: 80–3. National Botanical Institute, Pretoria.

Smith, C. A. 1966. *Common Names of South African Plants.* Government Printer. Pretoria.

Stafleu, F. A. & Cowan, R. S. 1976. *Taxonomic Literature* (7 vols.). Bohn, Scheltema & Haltema, Utrecht.

Van Jaarsveld, E. J. 1980. Succulent Lamiaceae of South Africa. *Aloe.* 18, 1 & 2: 33-5.

Van Jaarsveld, E. J. 1981. Suid-Afrikaanse *Plectranthus* spesies as tuinplante 1 & 2. *Veld & Flora.* 67, 1: 53-5 and 67, 2: 22-5.

Van Jaarsveld, E. J. 1982. Sukkulentagtige *Lamiaceae* van S. A. *Veld & Flora.* 68, 2: 43-7.

Van Jaarsveld, E. J. 1983. *Plectranthus hilliardiae* (Umtamvuna Spoorsalie): 'n Juweel van die Suidkus. *Veld & Flora.* 69, 2: 37-9.

Van Jaarsveld, E. J. 1987. *The Plectranthus handbook.* National Botanical Institute. Cape Town.

Van Jaarsveld, E. J. 1991. *Plectranthus fruticosus* and its cultivars with specific reference to cultivar Ellaphie. *Veld & Flora.* 77, 2: 42-3.

Van Jaarsveld, E. J. 1994. The genus *Plectranthus* and new cultivars. *Hortagrow.* 1: 16-18.

Van Jaarsveld, E. J. 1997. Notes on *Plectranthus* (*Lamiaceae*) from southern Africa. *Bothalia.* 27, 1: 1-6.

Van Jaarsveld, E. J. 2002. Lamiaceae. In Urs Eggli (ed.), *Illustrated Handbook of Succulent Plants.* Springer-Verlag, Berlin.

Van Jaarsveld, E. J. & Edwards, T. E. 1991. *Plectranthus reflexus. The Flowering Plants of Africa.* Vol. 51, 2: t 2034.

Van Jaarsveld, E. J. & Edwards, T. E. 2001. *Plectranthus hilliardiae* subsp. *australis,* a new taxon from Eastern Cape, South Africa. *Bothalia* 31, 1: 44-5.

Van Jaarsveld, E. J. & Hankey, A. 1997. *Plectranthus venteri,* a new species from the Northern Province, South Africa. *Aloe* 34, 1 & 2: 40-1.

Van Jaarsveld, E. J. & Van Wyk, A. E. 2004. *Plectranthus mzimvubuensis,* a new species from Eastern Cape, South Africa. *Bothalia* 34, 1: 30-2.

Van Wyk, A. E. & Smith, G. 2001. *Regions of Floristic Endemism in Southern Africa.* Umdaus Press, Pretoria.

Vegter, H. 1983. *Index Herbariorum,* part 2. The Hague, Netherlands.

Watt, J. M. & Breyer-Brandwijk, M. G. 1962. *The Medicinal and Poisonous Plants of Southern Africa,* 2nd ed. Livingstone, London.

Willis, J. C. 1973. *A Dictionary of the Flowering Plants and Ferns,* 8th ed. Cambridge.

Winter, P. J. D. & Van Jaarsveld, E. J. 2005. *Plectranthus porcatus,* a new species endemic to the Sekhukhuneland centre of plant endemism, Limpopo Province, South Africa. *Bothalia* 35, 2: 169-73.

List of synonyms

Synonym	Accepted name
Burnatastrum spicatum (E. Mey. ex Benth.) Briq.	***P. spicatus***
Coleus amboinicus Lour.	***P. amboinicus***
C. amboinicus var. *violaceus* Gürke	***P. amboinicus***
C. aromaticus Benth.	***P. amboinicus***
C. barbatus (Andr.) Benth.	***P. barbatus***
C. caninus (Roth) Vatke	***P. caninus***
C. carnosus Dinter	***P. neochilus***
C. comosus Hochst. ex Gürke	***P. ornatus***
C. crassifolius Benth.	***P. amboinicus***
C. daxo A. Shev.	***P. esculentus***
C. decumbens Gürke	***P. tetensis***
C. esculentus (N.E. Br.) G. Tayl.	***P. esculentus***
C. flavovirens Gürke	***P. caninus***
C. floribundus (N.E. Brown) Robyns	***P. esculentus***
C. forskohlii sensu Briq.	***P. barbatus***
C. melanocarpus (Gürke) Robyns & Lebrun	***P. tetragonus***
C. myrianthus (Briq.) Brennan	***P. hereroensis***
C. neochilus (Schltr.) Codd.	***P. neochilus***
C. omahekense Dinter	***P. caninus***
C. pentheri Gürke	***P. pentheri***
C. schinzii Gürke	***P. neochilus***
C. spicatus Benth.	***P. caninus***
C. tetensis Bak.	***P. tetensis***
C. vagatus E. A. Bruce	***P. tetensis***
Englerastrum floribundum (N.E. Brown) Th. Fries	***P. esculentus***
Englerastrum tetragonus (Gürke) Th. Fries jun.	***P. tetragonus***
Germanea cylindracea (Hochst. ex Benth.) Hiern	***P. cylindraceus***
Germanea laxiflora (Benth.) Hiern.	***P. laxiflorus***
Germanea urticifolia Lam.	***P. fruticosus***
Majana amboinica (Lour.) Kuntze	***P. amboinicus***
Neomuellera damarensis S. Moore	***P. hereroensis***
Ocimum hadiense Forssk.	***P. hadiensis* var. *hadiensis***
Orthosiphon ambiguus	***P. ambiguus***
Plectranthus aromaticus Roxb.	***P. amboinicus***
P. arthropodus Briq.	***P. fruticosus***
P. aurifer Dinter ex Launert	***P. hereroensis***
P. behrii Compton.	***P. fruticosus***
P. biflorus Bak.	***P. tetragonus***
P. calycinus var. *pachstachyus* (Briq.) T. Cooke	***Rabdosiella calycina****
P. charianthus Briq.	***P. fruticosus***
P. coloratus E. Mey. ex Benth.	***P. ambiguus***
P. cooperi T. Cooke	***P. grallatus, P. dolichopodus***
P. densiflorus T. Cooke	***P. cylindraceus***
P. draconis Briq.	***P. woodii***
P. dregei Codd.	***P. ambiguus***
P. elegantulus Briq.	***P. ciliatus***
P. floribundus N.E. Brown	***P. esculentus***
P. floribundus var. *longipes* N.E. Brown	***P. esculentus***
P. forskohlaei sensu Ait.	***P. barbatus***
P. forskohlaei Vahl	***P. hadiensis* var. *hadiensis***
P. galpinii Schltr.	***P. fruticosus***
P. hadiensis var. woodii (Gürke) Codd	***P. woodii***
P. hylophilus sensu Cooke	***P. laxiflorus***
P. krookii Gürke ex Zahlbr.	***P. grallatus***
P. kuntzianus Domin.	***P. strigosus***
P. madagascariensis var. *aliciae* Codd	***P. aliciae***
P. madagascar var. *ramosior* Benth.	***P. ramosior***
P. marrubioides Hochst. ex Benth.	***P. cylindraceus***
P. matabelensis Bak.	***P. hereroensis***
P. melanocarpus Gürke	***P. tetragonus***
P. moschosmoides Bak.	***P. cylindraceus***
P. myrianthus Briq.	***P. hereroensis***
P. natalensis forma *glandulosa* Phillips	***P. grallatus***
P. natalensis Gürke	***P. ciliatus***
P. otaviensis Dinter	***P. hereroensis***
P. pachyphyllus Gürke ex T. Cooke	***P. hadiensis* var. *hadiensis***
P. pachystachyus Briq.	***Rabdosiella calycina****
P. parviflorus Gürke	***P. strigosus***
P. peglerae T. Cooke	***P. fruticosus***
P. praetervissus Briq.	***P. grallatus***
P. pyramidatus Gürke	***Rabdosiella calycina****
P. spiciformis R. A. Dyer	***P. cylindraceus***
P. strigosus var. *lucidus* Benth.	***P. lucidus***
P. subspicatus Hochst.	***P. spicatus***
P. tomentosus Benth.	***P. hadiensis* var. *tomentosus***
P. transvaalensis Briq. var. *grandifolia* T. Cooke	***P. grallatus***
P. urticifolius (Lam.) Salsb.	***P. fruticosus***
P. vagatus (E. A. Bruce) Codd	***P. tetensis***
P. villosus T. Cooke	***P. cylindraceus***
P. zatarhendi var. *tomentosus* (Benth.) Codd	***P. hadiensis* var. *tomentosus***
P. zatarhendi var. *woodii* (Gürke) Codd	***P. woodii***

* Transferred to *Rabdosiella* by Dr L. Codd in *Bothalia* 15:9 (1984).

Glossary

abaxial surface the lower surface (e.g. of a leaf), facing away from the axis or stem

accrescent increasing in size

acuminate tapering to a fine point

adaxial upper surface, facing the axis or stem

alternate spirally arranged parts (not opposite)

apiculus a short sharp point

arborescent having tree-like characteristics

axillary growing from the axil

bifid forked or divided into two

bract modified leaf between the calyx and the normal leaves

caducous dropping early

calyx outer whorl of floral leaves (sepals), protecting the young flower in the bud; in the case of *Plectranthus*, the fused sepals which form a tube

caudiciform resembling a caudex (= a rootstock or trunk, referring in succulents to the swollen basal portion of a stem or rootstock). In *P. ernstii*, 'caudiciform' refers to the trunk-like basal part of the stem

chartaceous papery

circinate coiled into a ring, like a watch-spring

circumscissile dehiscing circularly, as at the calyx base in *Aeollanthus*

coma a tuft of leaves or bracts crowded together like the crown of a palm. In the subgenus *Calceolanthus*, *P. neochilus* has a distinctive coma of neatly arranged large bracts protecting the young inflorescence

conduplicate folded together lengthwise

cordate heart-shaped

corolla all the floral parts excluding the calyx. In *Plectranthus*, refers to the petals which are fused to form a tube

cremnophilous cliff-dwelling

crenate with rounded teeth

cultivar a cultivated form of a plant

cuneate wedge-shaped

cymbiform boat-shaped

cyme an inflorescence in which the terminal bud ends in the oldest flower, and the lateral branches end in progressively younger flowers

Glossary cont.

deciduous of an organ that abscisses, often during the dry season
decumbent lying flat with the end curving upwards
decurrent of a leaf where the blade extends down the petiole or stem
decussate with each opposite pair of leaves (or flowers) at right angles to the next
deflexed bent abruptly downwards
dentate with a toothed margin
denticulate with a small-toothed margin
dichasium a cymose inflorescence with each branch giving rise to two other branches
didynamous having two stamens longer than the rest (as in *Plectranthus zuluensis*)
eglandular without glands
elliptic tapering at each end with the widest point in the centre
emarginate notched at the abrupt or rounded apex
endemic confined to a specific area
entire continuous and not broken into by teeth or lobes (often of margins)
epidermis outer cell layer
filament thread-like stalk of the anther
geniculate bent abruptly lie a knee
gibbose swollen, more so in one part than another. In the section *Coleoides*, the mature calyx is characteristically swollen at the base; this is a useful feature distinguishing plants in this section from those in section *Plectranthus*
glabrescent becoming hairless at maturity
gland dot minute attractive glands found on the surface of *Plectranthus* leaves (mainly on the undersides), stems and sometimes flowers. They can be of various shapes and colours, and are occasionally translucent
globose almost spherical
glomerate collected in a head
hispidulous bearing rough hairs to a lesser degree
iconotype an illustration serving as the type for a taxon when no actual specimen was cited
imbricate overlapping, like tiles on a roof
indumentum a hairy covering
introgression spreading of genes due to hybridization
isotype duplicate of the type specimen
lanceolate spear-shaped
lectotype the specimen or illustration selected as the type specimen by a subsequent taxonomist, when the original author used more than one specimen in his description of a taxon
linear parallel-sided and much longer than broad
lithophyte a plant that grows on rocks, stones or boulders
mucronate with a short sharp point, mainly situated at the leaf tips
neotype the specimen or illustration selected when no specimens or illustrations were designated by the original author
oblong longer than broad, with sides more or less parallel
obtuse an angle greater than 90 degrees
ovary the female part of a flower, the basal carpel carrying the ovules which later will turn into seed
ovate egg-shaped, broadest below the middle
panicle branched raceme with stalked flowers
parenchymatous tissue spherical storage cells
pedicel stalk of single flower
pedicellate stalked, mainly referring to the flower or fruit
peduncle stalk of a flower or inflorescence
pendulous hanging, suspended
petiole leaf-stalk
pilose with simple stiff hairs
procumbent growing horizontally, and supported by the ground
proximal towards the point of attachment
pseudoraceme resembles a raceme, but isn't one
pseudostem resembles a stem, but isn't one
puberulous slightly pubescent
pubescent hairy
punctate marked with dots
raceme inflorescence with a lengthened axis bearing stalked flowers, the oldest flower at the base
rachis the axis of an inflorescence
reticulate with a network of, for example, veins on the undersurface of some leaves
retuse notched at the abrupt or obtuse apex
rubropunctate carrying red points or dots
scabrid rough to the touch
scandent climbing, though with a tendency to lean
scorpioid cyme a coiled inflorescence (like a scorpion's tail) with flowers alternating in two opposite ranks
sessile without a stalk
sigmoid double curved and in opposite directions
soboliferous bearing vigorous shoots or stolons, usually from below ground
staminode a infertile stamen, usually reduced
stigma the end of the style, receptive to pollen
stipitate stalked
stipule basal appendage of a leaf, often with a protective function for young buds
stolon a basal sucker or runner, capable of forming a new plant
striate marked with parallel lines
strigose short rough hairs pointed in the same direction, as in the leaves of *Plectranthus strigosus*
style upper part of the pistil, between the ovary and stigma
subequal almost equal
subimbricate almost imbricate
subrotund almost rounded
subspicate almost spike-like
subulate awl-shaped
succulent plant from arid habitats with thickened water-storing roots, leaves or stems
suffrutex shrublet or shrub with branches dying back after flowering to a persistent rootstock
sympatric of species that occur together
taxonomy classification
terete cylindrical
tomentose covered with cottony hairs
tomentulose slightly tomentose
trichome a hair-like outgrowth of the epidermis
truncate abruptly ending
type specimen the herbarium specimen used for the description of a species
variety subdivision of a species, recognizing the natural variation of a group of individuals within a species (or subspecies) not geographically isolated
versatile (of an anther) turning freely on the apex of the filament
verticillaster a false whorl made up of decussate cymes
verticils whorls of flowers, or a circular arrangement of flowers
villous covered in long weak hairs
viscid sticky

Index

Numbers in *italics* refer to illustrations. Numbers in **bold** refer to main entries. Accepted *Plectranthus* names in ***bold italics,*** synonyms or hybrids in *italics*